Valley of Grass

Tallgrass Prairie and Parkland
of the Red River Region

Erik,

Your steadfast and enlightened determination will serve you well.

I have enjoyed working with you, especially this summer. Thank you for all your excellent work!

sincerely

H. Paul

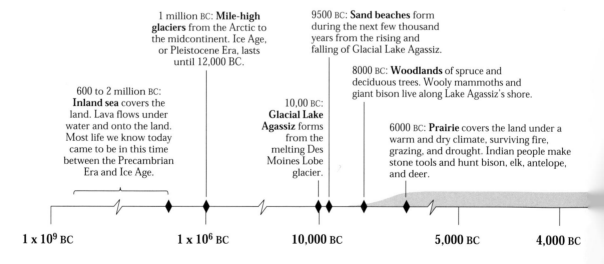

1 million BC: **Mile-high glaciers** from the Arctic to the midcontinent. Ice Age, or Pleistocene Era, lasts until 12,000 BC.

9500 BC: **Sand beaches** form during the next few thousand years from the rising and falling of Glacial Lake Agassiz.

8000 BC: **Woodlands** of spruce and deciduous trees. Wooly mammoths and giant bison live along Lake Agassiz's shore.

600 to 2 million BC: **Inland sea** covers the land. Lava flows under water and onto the land. Most life we know today came to be in this time between the Precambrian Era and Ice Age.

10,00 BC: **Glacial Lake Agassiz** forms from the melting Des Moines Lobe glacier.

6000 BC: **Prairie** covers the land under a warm and dry climate, surviving fire, grazing, and drought. Indian people make stone tools and hunt bison, elk, antelope, and deer.

1 x 10⁹ BC   1 x 10⁶ BC   10,000 BC   5,000 BC   4,000 BC

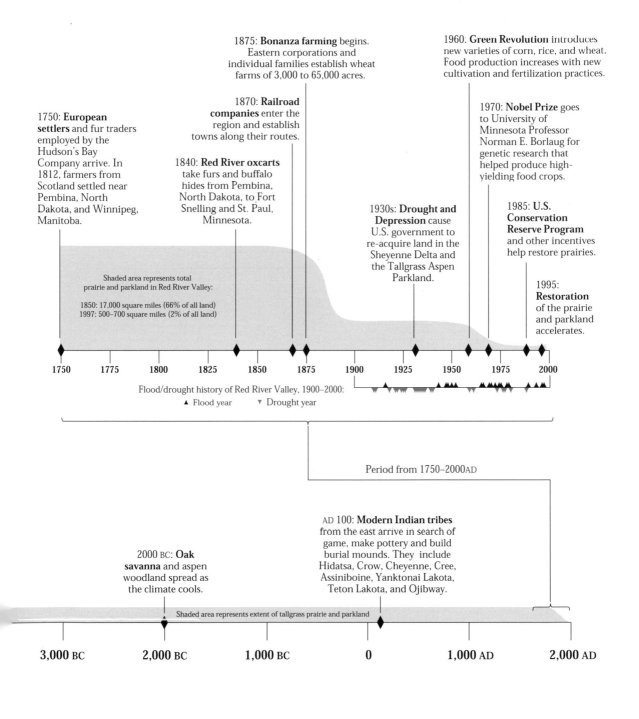

1875: **Bonanza farming** begins. Eastern corporations and individual families establish wheat farms of 3,000 to 65,000 acres.

1960. **Green Revolution** introduces new varieties of corn, rice, and wheat. Food production increases with new cultivation and fertilization practices.

1870: **Railroad companies** enter the region and establish towns along their routes.

1750: **European settlers** and fur traders employed by the Hudson's Bay Company arrive. In 1812, farmers from Scotland settled near Pembina, North Dakota, and Winnipeg, Manitoba.

1840: **Red River oxcarts** take furs and buffalo hides from Pembina, North Dakota, to Fort Snelling and St. Paul, Minnesota.

1970: **Nobel Prize** goes to University of Minnesota Professor Norman E. Borlaug for genetic research that helped produce high-yielding food crops.

1930s: **Drought and Depression** cause U.S. government to re-acquire land in the Sheyenne Delta and the Tallgrass Aspen Parkland.

1985: **U.S. Conservation Reserve Program** and other incentives help restore prairies.

1995: **Restoration** of the prairie and parkland accelerates.

Shaded area represents total prairie and parkland in Red River Valley:

1850: 17,000 square miles (66% of all land)
1997: 500–700 square miles (2% of all land)

1750   1775   1800   1825   1850   1875   1900   1925   1950   1975   2000

Flood/drought history of Red River Valley, 1900–2000:
▲ Flood year   ▼ Drought year

Period from 1750–2000AD

AD 100: **Modern Indian tribes** from the east arrive in search of game, make pottery and build burial mounds. They include Hidatsa, Crow, Cheyenne, Cree, Assiniboine, Yanktonai Lakota, Teton Lakota, and Ojibway.

2000 BC: **Oak savanna** and aspen woodland spread as the climate cools.

Shaded area represents extent of tallgrass prairie and parkland

3,000 BC   2,000 BC   1,000 BC   0   1,000 AD   2,000 AD

VALLEY OF GRASS

Tallgrass Prairie and Parkland of the Red River Region

Kim Alan Chapman
Adelheid Fischer
Mary Kinsella Ziegenhagen

North Star Press of St. Cloud, Inc.

Library of Congress Cataloging-in-Publication Data

Chapman, Kim, 1955-
    Valley of grass : tallgrass prairie and parkland of the Red
River region / Kim Chapman, Adelheid Fischer, Mary Kinsella
Ziegenhagen.
        144 p. 26 cm.
        Includes bibliographical references (p. 119).
        ISBN: 0-87839-127-4 (pbk. : alk. paper)
        1. Prairie ecology—Red River Valley (Minn. and N.D.-Man.)
2. Prairie conservation—Red River Valley (Minn. and N.D.-
Man.) 3. Red River Valley (Minn. and N.D.-Man.) I. Fischer,
Adelheide. II. Ziegenhagen, Mary. III. Title.
QH104.5.R33C48 1998
577.4'4'097841--dc21   98-19433
                                                                        CIP

Front and back cover design and photographs © Richard Hamilton
Smith Photography. Creation of "Valley of Grass" portion of the
title credited to Richard Hamilton Smith Photography.

All maps and illustrations, unless noted, by Patricia Isaacs of Parrot
Graphics.

Copy editing by Mary Keirstead.

ISBN: 0-87839-127-4

Printed in Canada by Friesens Corporation in the Red River Valley.

Published by North Star Press of St. Cloud, Inc., P.O. Box 451,
St. Cloud, Minnesota 56302, in conjunction with The Nature
Conservancy.

Acknowledgments

We thank those individuals who we interviewed or who provided information that became the content of this book. They gave generously of their time and ideas, and passed on to us an enthusiasm for their subject and livelihoods. We wish to acknowledge the following people for their substantial contribution in person, in writing, or through the materials they sent to us. (Individuals with an asterisk also reviewed portions or all of the book.)

North Dakota

Bob Barker, North Dakota State University, Fargo; John Challey, biologist, Fargo*; Gary Clambey, North Dakota State University, Fargo; Gerald Fauske, North Dakota State University, Fargo; David Hopkins, North Dakota State University, Fargo*; Jay Mar, Lake Agassiz Resource Conservation and Development Council, Fargo; Marcia McMullen, North Dakota State University, Fargo; Mikkel Pates, Fargo Forum, Fargo; Jimmie Richardson, North Dakota State University, Fargo*; David Rider, North Dakota State University, Fargo; Steve Schumacher, U.S. Forest Service, Lisbon*; Dennis Sexhus, North American Bison Cooperative, New Rockford; John Sheppard, Economic Development Commission, Mayville; Tom Spiekermeier, farmer, Sheldon; Bea Wall, rancher, Sheldon; Larry Woodbury, rancher, McLeod

Minnesota

Luther Aandlund, Minnesota Department of Natural Resources, Fergus Falls; Deborah Allan, University of Minnesota, St. Paul*; Maggie Alms, Blue Earth Environmental, Lake Crystal; Bill Berg, Minnesota Department of Natural Resources, Grand Rapids; Brad Bjerken, rancher, Felton; George Boody, Land Stewardship Project, White Bear Lake*; Cindy Buttleman, Minnesota Department of Natural Resources, Bemidji; Tom Clark, Minnesota Department of Revenue, St. Paul; Robert Dana, Minnesota Department of Natural Resources, St. Paul*; Mark Davis, Macalester College, St. Paul; John Evert, farmer and county commissioner, Moorhead; Don Faber-Langendoen, The Nature Conservancy, Minneapolis; Russell Fitzgerald, retired railroad worker, Dilworth; Art Fosse, designer and builder of Rothsay Prairie Chicken, Rothsay; Elmer Fuchs, rancher, Felton; Joe Gartner, Moorhead Regional Science Center, Glyndon; Wayne Goeken, Agassiz Environmental Learning Center, Fertile; Tim Hagle, farmer, Red Lake Falls; Mark Hanson, Minnesota Department of Natural Resources, Bemidji; Fred Harris, Minnesota Department of Natural Resources, St. Paul*; Ross Hier, Minnesota Department of Natural Resources, Crookston*; Paul Kaste, seed farmer, Fertile; Tim Magnuson, Clay County government, Moorhead; Jerry Nagel, Northern Great Plains Rural Development Commission, Crookston; Kirt Nigard, Extension Agent, Kittson County; Wendell Olson, U.S. Fish and Wildlife Service, Fergus Falls; Ralph and Roberta Ouse, retired farmers, Rothsay; Mark Peihl, Clay County Historical Society, Moorhead; Richard Pemble, Moorhead State University, Moorhead; Catherine Reed, University of Minnesota, St. Paul*; Donna Stockrahm, Moorhead State University, Moorhead*; Dan Svedarsky, University of Minnesota, Crookston*; Steve Taff, University of Minnesota, St. Paul*; Dave Tilman, University of Minnesota, St. Paul*; Gerald Van Amburg, Concordia College, Moorhead*; Terry Wolfe, Minnesota Department of Natural Resources, Crookston*

Manitoba

Wayne Arseny, Mayor, Emerson; Lorne Colpitts, Manitoba Habitat Heritage Corporation, Winnipeg; Murray Gillespie, Manitoba Department of Natural Resources, Winnipeg; Jason Greenall, Manitoba Department of Natural Resources, Winnipeg*; Bob Jones, Manitoba Department of Natural Resources, Winnipeg*; Diane Kunec, Manitoba Department of Natural Resources, Winnipeg Marilyn Latta, Manitoba Naturalists Society, Winnipeg; Janet Moore, Critical Wildlife Habitat Program, Winnipeg; Bob Oleson, Manitoba Provincial Government, Winnipeg*; Doug Pastuck, Manitoba Department of Natural Resources, Winnipeg; Linda Shewchuk, Ukrainian Museum and Village Society, Gardenton; Glen Suggett, Manitoba Department of Natural Resources, Winnipeg

Other States and Provinces

Steve Apfelbaum, Applied Ecological Services, Brodhead, Wisconsin*; Bob Hamilton, The Nature Conservancy, Pawhuska, Oklahoma; Mark Leach, University of Wisconsin, Madison, Wisconsin; Dennis Schlicht, teacher and lepidopterist, Center City, Iowa*; Gerald Selby, The Nature Conservancy, Des Moines, Iowa; John Sidle, U.S. Forest Service, Chadron, Nebraska*; Al Steuter, The Nature Conservancy, Ainsworth, Nebraska*; Larry Tieszen, Augustana College, Sioux Falls, South Dakota; John Toepfer, prairie chicken researcher, Stevens Point, Wisconsin; David Wedin, University of Toronto, Toronto, Ontario

The authors especially wish to thank the team of individuals who advised us from the beginning of this project, and who dedicated their energy and imagination to making this book possible.

Peter Buesseler, Minnesota Department of Natural Resources, Fergus Falls, Minnesota*; Jim Erkel, The Nature Conservancy, Minneapolis; Gene Fortney, Nature Conservancy of Canada, Winnipeg*; Nelson French, The Nature Conservancy, Minneapolis; Gabrielle Horner, The Nature Conservancy, Minneapolis*; Wayne Ostlie, The Nature Conservancy, Minneapolis; Joe Satrom, The Nature Conservancy, Bismarck*; Andy Schollett, The Nature Conservancy, Bismarck; Carol Scott, Manitoba Department of Natural Resources, Winnipeg*; Brian Winter, The Nature Conservancy, Glyndon, Minnesota.*

The authors heartily thank Kendra McLaughlan, University of Minnesota doctoral student, for her diligent research and intelligent interviews with people in and around the Red River region, which provided so much information for this book.

The strength of an organization is its people, and The Nature Conservancy staff in the Dakotas and Minnesota are exceptional. With their energy, intelligence, and knowledge, they provided the authors the needed support and assistance that allowed them to complete the book. Mary Waterhouse provided considerable organizational help to all of us during the project, which included typing sections of the book. Julie Muehlberg applied her artistic eye and design skills in the creation and selection of graphics and photographs. The development team in Minnesota, especially Gloria Karbo, Peggy Ladner, and Jessica Pihlstrom, found the resources to pay for the writing, production, and printing of the book. Financial support was provided by the Carolyn Foundation, Mrs. Ellie Sturgis, Mrs. Reuel D. Harmon, and the James Ford Bell Foundation, without whose generosity this book would not have come to be.

Foreword

Although this book is wonderfully full of details about how the prairies and parklands of the Red River Valley function—more information of this kind than has ever before been gathered in one place—and although these details are vital to the understanding of a notably rich and beautiful place, the importance of this book lies not so much in its particulars as in its attitude. The authors suppose that there need be no unresolvable conflict between the conservation of natural resources and the building of healthy human communities. Indeed, they suggest, the two endeavors can be complementary.

This is something new, particularly in farming country. A lot of energy and thought has been put into the problem of protecting wild places from agriculture, but very little has been invested in thinking about how conservationists and farmers might be allies, to their mutual benefit.

Farmers depend in many ways upon healthy natural systems: for the structure and fertility of soils, for erosion control, for the recycling of nutrients, for stable climates, for the services of wild pollinators of plants, for assistance in battling pests and pathogens, and for the recharging of groundwater supplies, among others. It is easier and more profitable to farm in a healthy landscape than in the midst of one that is sickly and barren.

At the same time, the goal of preserving wild species and of maintaining whole ecosystems cannot be accomplished on conservation preserves alone. The tallgrass prairie system, in particular, is already so fragmented and so nearly destroyed that, even if every square inch of what remains were to be set aside forever, we could not be confident of saving prairie resources for the future. Only when conservationists join hands with farmers, understanding their goals to be mutually beneficial, will the health of prairie landscapes begin to improve.

The key to this alliance lies in seeing the prairie not so much as a set of resources—although it is very much that—but as a teacher. For about as long as there has been agriculture, there have also been tallgrass prairies. Over thousands of years, prairie plants and critters have developed highly successful strategies for surviving and prospering in just such conditions of land and climate as we find in the Red River Valley. The prairie "knows," in a deep sense, a great many things about persevering in this landscape that we can as yet hardly even guess at. It contains far more information than we have yet extracted from it. There is no telling what the prairie might teach us about how to make long-lasting communities here.

The strongest link human communities have to wild ones is that human communities depend upon the kinds of information that only wild communities can convey. Wild places are our greatest libraries; they lie at the heart, therefore, of human civilizations, which have been constructed on the foundation of the peculiarly human need to know. We will not only prosper most, but be at our most civilized when we see nature first not as something to be used but as something to be learned from.

This book offers that vision and begins to suggest practical ways in which it might be carried forward. It is, on that account, one of the most hopeful pieces of writing I have seen in years.

Paul Gruchow
Moorhead, Minnesota
February 1998

Table of Contents

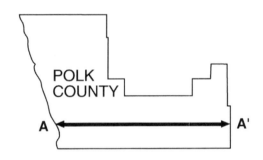

POLK COUNTY

A ◄──────────────────────► A'

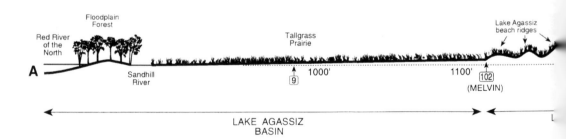

Red River of the North

Floodplain Forest

Sandhill River

Tallgrass Prairie

1000'

⑨

1100'

Lake Agassiz beach ridges

102
(MELVIN)

A

◄──► ◄──
LAKE AGASSIZ
BASIN

East-West Transect across Polk County
through the Center of T148N
Showing Original Vegetation and Major Geologic Features
1:150,000

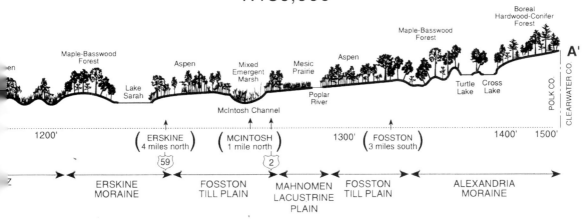

Illustration by Tom Klein. (Courtesy of the State of Minnesota Natural Heritage and Nongame Research Program and County Biological Survey)

Valley of Grass

Tallgrass Prairie and Parkland
of the Red River Region

Chapter 1

Our Common Ground
The Ties That Bind

These are the scenes which must be witnessed and felt before the mind forms a true conception of the Red River prairies in that unrelieved immensity which belongs to them in common with the ocean, but which, unlike the changing and unstable sea, seem to promise a bountiful recompense to millions of our fellow-men.

Henry Youle Hind,
leader of the Red River Exploring Expedition,
Narrative of the Canadian Red River Exploring Expedition of 1857

There was an edge to the wind, and in the darkening days of autumn, the prairie grasses glowed wine red and gold. Even though winter would soon sweep in from the north, Matthew Campbell walked the land and was filled with hope and gratitude. He was among the thousands of land-seekers who had journeyed west to the nation's heartland in search of crystalline springs and tree-lined rivers, wildlife free for the taking, and soil so dark and fertile it glistened in a settler's fist like nuggets of black gold. On October 28, 1872, Campbell wrote to his sister about his newfound fortune from a camp near the Buffalo River east of what is now Moorhead, Minnesota.

Wheat harvest on the Fritz Gruhl farm, Cromwell Township, Clay County, Minnesota, ca. 1905. (Courtesy of the Flaten/Wange Collection, Clay County Historical Society)

1

Dear Sister Isabella,

You ask whether this country will produce vegetables. Just think of the nicest potatoes you ever saw and then multiply by two and you will have the answer. Prairie chickens, ducks, and wild geese are abundant. One day I shot two chickens with one shoot, two large ducks at another, and three smaller ones at another. I saw a couple of elk the other day. There are wolves and badgers and a few bears. Wild plums are plenty and excellent.

I think this will be a good country. The Lord has sent me here, and I think I will soon do as well here, if not better, than I have ever done. I am homesteading. Our children are remarkably healthy. . . . I feel myself as if my own age is renewing. The landscape view is sublime. The mirage is wonderful. Soon the sun dogs will walk their circuits from the sun, scenes which other climates do not afford.

> Your Brother,
> Matthew Campbell

Threshing wheat near Hawley, Clay County, Minnesota, ca. 1910. (Courtesy of the Flaten/Wange Collection, Clay County Historical Society)

2

Today, the Red River Valley earth that rewarded Matthew Campbell's family with bushels of vegetables and wild plums has become the breadbasket of the world. Red River wheat, beet sugar, and potatoes are staples of the global larder.

To read the story of the Red River Valley solely in terms of its agricultural prowess, however, is to overlook the other enduring ties that bind people to the land. As the Campbells and the generations that followed worked the land, they built a community in place—not just any place. Here, the rhythms of labor and play, the occasional celebrations, and the ordinary details of daily life grew out of the land's ebb and flow. Today, as they did more than a century ago, people continue to find a place that measures up to their feelings of hope and renewal. Then, as now, the source of this hope and renewal is the prairies and parkland.

The Red River Valley was once the bed of North America's largest inland lake, known as Glacial Lake Agassiz. Over thousands of years, a glacier bore down from the north, retreating and returning, shaping and reshaping beaches and ridgelines with its melting waters, building basins where waters pooled, and cutting spillways that drained the land in a labyrinth of streams.

After the lake waters receded, tallgrass prairie slowly took hold in the valley. Here, at the rate of about an inch each hundred years, a rich, dark soil accumulated to as much as five feet deep.

The Red River region was part of the great North American tallgrass prairie, a vast grassland that knit the continent together in a plush, green seam from Texas north to Manitoba. Today, this ecosystem, whose towering grasses could swallow travelers on horseback for days at a time, is all but extinct.

On the edges of the Red River Valley, however, some of the largest and best examples of tallgrass prairie and parkland still survive. Amid the deltas and beaches formed by ancient Lake Agassiz survive an array of habitats that range from rare fens and wet meadows to dry, sandy grasslands and aspen-oak savannas. Rare orchids, butterflies, and birds thrive here as they have for thousands of years. These

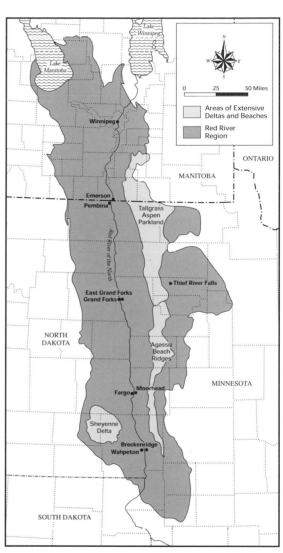

Ancient deltas and beaches of Glacial Lake Agassiz. (Source: The Nature Conservancy Great Plains Program)

3

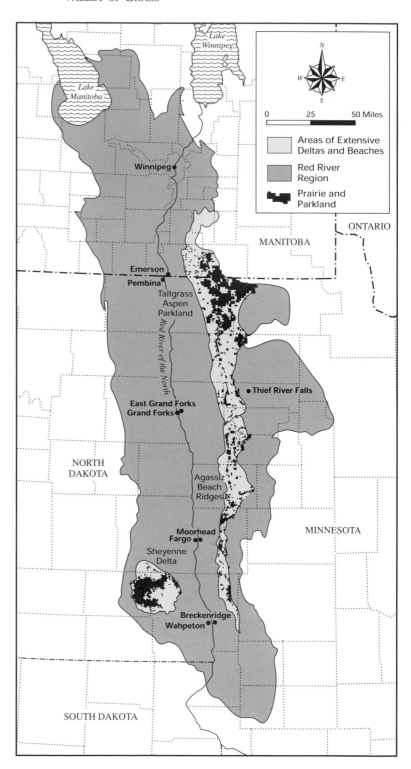

The remains of tallgrass prairies and parkland are concentrated at the edges of the Red River Valley, on ancient deltas and beaches of Glacial Lake Agassiz. (Source: The Nature Conservancy Great Plains Program, Manitoba Conservation Data Centre, Minnesota Natural Heritage and Nongame Research Program and North Dakota Natural Heritage Program)

landscapes include the Sheyenne Delta, located in southeastern North Dakota; the Tallgrass Aspen Parkland, which reaches from Winnipeg in southern Manitoba into Marshall County, Minnesota; and the Agassiz Beach Ridges, which extend from near Rothsay to beyond Crookston, Minnesota.

These places cannot survive, however, without ingenuity and care. To save the remaining tallgrass prairies and parkland, a balance between their use and conservation must be struck. This book is about re-visioning the relationship to the land. It is about regarding the land as a classroom for human history as much as a classroom for natural history. It is about renewing spirits as much as diversifying local economies. It is about thrilling to the booming of prairie chickens as much as harvesting medicines to heal disease. It is about envisioning a laboratory for agricultural research as much as providing a living textbook for schoolchildren. It is about preserving aquifers and cleansing urban stormwater as much as preserving a place where the song of a meadowlark keeps company with the solitude of the sky.

To ensure that natural lands will be here to meet these complex needs, many public and private groups are looking beyond state lines

A dickcissel with black throat and yellow chest sings his territorial song from a perch in a prairie. (The Nature Conservancy photo archives)

Space for Chickens?

Greater prairie chicken. (Photo by Dominique Braud)

The story of the greater prairie chickens in the Red River region is a story of ups and downs. Every ten years or so, the population peaks, and five years later it bottoms out. Before 1870, there were sharp-tailed grouse but few prairie chickens in the prairies and parkland. Breaking the prairie for crop farming drove out the sharp-tails and attracted the chickens. Nearly 420,000 were shot in 1925 alone. Following a decline, the last open season on prairie chickens was 1942. By 1965 they had largely stopped breeding outside the Red River region, in spite of thousands of acres of prairies. There was simply too much cropland relative to the small and separate prairies and wetlands needed for nesting and winter survival.

The United States Conservation Reserve Program (CRP) turned out to be a boon to prairie chickens, producing in 1992 the highest numbers since surveys began. Researchers speculate that without CRP lands, the wet springs of 1992 and 1993 might have driven the prairie chicken to lower numbers. For a dynamic species like the prairie chicken, there is a choice: use the big prairie well and enjoy its exuberant strutting and thrumming calls for a thousand future springs.

[Theodore Roosevelt and his brother, Elliott,] killed 404 animals on a twenty-four-day hunting trip [to Moorhead, Minnesota, in 1880]. Two-hundred eight were shot in eight days in Clay County, including seventy sharp-tails, seventy-seven prairie chickens, twelve plovers, thirty-seven ducks, two jackrabbits, two geese and assorted doves, coots and grebes. That represents a tremendous slaughter in today's standards, but it was not unusual for the time.

Mark Peihl, archivist, Clay County Historical Society,
Clay County Historical Society Newsletter, September/October 1992

Making Room for Chickens

Careful use of the prairie landscape creates conditions that are right for the greater prairie chicken. Over 4,000 of the birds live in the Agassiz Beach Ridges because of its balanced mix of big prairies, cropland, and hay meadows or pasture. Dr. Dan Svedarsky at the University of Minnesota-Crookston has studied the prairie chicken for over twenty years. He and other researchers suggest that the best habitat is found where a third to half the land is native prairie or tame grass, a quarter is sedge meadow and lowland brush, and less than a tenth is trees. The rest can be cropland. Bigger prairies attract more prairie chickens year after year than smaller ones in a similar setting, and prairie chickens avoid areas with dense trees. About a quarter of the prairies and pastures should be burned, mowed or grazed every year.

In 1996, Svedarsky and Gerry Van Amburg of Concordia College completed a study of the Sheyenne National Grasslands. Here they found that the greater prairie chicken was not as common as it might be. The quality of both forage and prairie chicken habitat could be increased, they concluded, by grazing in a way that favors native plants over non-native plants, by more prescribed burning of prairies and lowlands, and by leaving denser grass cover where prairie chickens are common. If people who live and work in the area can work out the details of how to do this, Svedarsky and Van Amburg think that cattle and chickens can get along together nicely.

Dan Svedarsky. (Photo by Michael Burian)

Greater prairie chickens dancing on booming ground. (Photo by Brian Winter)

and international boundaries to connect and enlarge remnant clusters of prairies and parkland with green corridors, and to manage and restore existing habitat complexes: for example, the Tallgrass Prairie Preserve, near Tolstoi, Manitoba; the Bluestem-Margherita Prairie, east of Fargo-Moorhead; and the Sheyenne Grasslands and Delta, between Valley City and Wahpeton, North Dakota.

Research has shown that large habitat patches with buffers and connecting corridors, which allow plants and animals to move from place to place, form more robust and long-lasting habitats. Isolated colonies of plants and animals in small habitats either disappear or lose genetic diversity over time. Prairies and parkland of at least 2,000 acres are far better at surviving extreme drought and other climate changes because drought-resistant plants are more numerous and varied, providing more habitat for birds and other animals. Furthermore, when a large prairie is invaded by non-native grasses and animals or harmed by pollution, the vegetation around the edges can serve as a buffer to screen out aggressors and protect the core habitat.

But saving the tallgrass prairie and parkland is not just about saving the unique assemblage of plants and animals that evolved in

More species of animals and plants will thrive in larger prairies than in smaller ones. In small prairies, many kinds of mammals and birds do not find enough of the right kind of food, shelter, and living space to raise their young and survive. (Acres needed are approximate. Illustration by Elizabeth Longhurst)

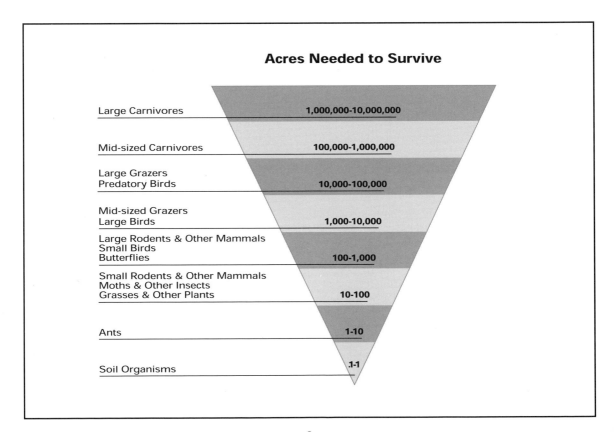

Acres Needed to Survive

Large Carnivores	1,000,000-10,000,000
Mid-sized Carnivores	100,000-1,000,000
Large Grazers / Predatory Birds	10,000-100,000
Mid-sized Grazers / Large Birds	1,000-10,000
Large Rodents & Other Mammals / Small Birds / Butterflies	100-1,000
Small Rodents & Other Mammals / Moths & Other Insects / Grasses & Other Plants	10-100
Ants	1-10
Soil Organisms	.1-1

this ancient matrix of beauty and abundance. It is also about honoring human ties to the land. This book profiles the efforts of people from North Dakota, Manitoba, and Minnesota who, like the generations before them, are finding opportunities for a rich life in the tallgrass prairie and parkland.

In only a few places in North America can the tallgrass prairie and parkland be preserved and restored as it existed for millennia, where the land is spacious, where the habitat remnants are large, and where most of the essential animals and plants continue to live—places where generations of people toiled to claim a fruitful land and found it worthy of their best efforts. The Red River region is such a place.

As bison did for hundreds of years, children cavort against the rubbing rock at Blazing Star Prairie, Clay County. (The Nature Conservancy photo archives)

9

(Photo by Tom Spiekermeier)

The Sheyenne Delta

One of thirty-three deltas formed on the edges of Glacial Lake Agassiz, the Sheyenne is second in size only to the Assiniboine Delta in Manitoba. This 450,000-acre sandy area concentrated in Ransom and Richland Counties, North Dakota, is characterized by dunelike formations shaped by the wind before native grasses stabilized the landscape. The soils are productive in spite of the region's history of drought and wind erosion. Most of the 127,000 acres of intact native habitat exist in a single immense block. Bur-oak savannas grow in river valleys and dune blowouts. Large concentrations of the western prairie fringed orchid, a nationally threatened plant, make the delta a prime research site for this rarity and its moth pollinators. The uncommon, handsome sedge occurs here, as well as the elusive Dakota skipper butterfly, and one of two reproducing populations of greater prairie chickens in North Dakota, besides the one near Grand Forks.

(Photo by Robert Dana)

Tallgrass Aspen Parkland

Taking its name from the open, parklike appearance of grasslands loosely dotted by shrub thickets and groves of trees, the 1,214,000-acre Tallgrass Aspen Parkland landscape, as seen from the air, is a mosaic of sedge meadows and brush prairies, wetlands and sand dunes, willow swamps and aspen groves surrounded by agriculture. This region straddles the United States-Canadian border. Minnesotans call these habitats collectively the aspen parkland, while Manitobans call the open habitats tallgrass prairie. To acknowledge these two traditional names and to describe the unique mingling here of two ecosystems—the tallgrass prairie and aspen parkland—this book refers to the region as the Tallgrass Aspen Parkland and refers to the habitats that characterize this region as tallgrass aspen parkland.

Over 225,000 acres of native habitat have been identified to date, with thirteen large areas totaling over 150,000 acres. In this vast constellation of natural systems, tallgrass prairie species intermingle with plants and wildlife of the aspen woodlands and northern peatlands. Bears, wolves, elk, and moose inhabit the land, and sharp-tailed grouse, whose numbers continue to drop (fifty percent decline in Minnesota in the early 1990s), find refuge in this landscape.

11

Agassiz Beach Ridges

These beaches were formed by the waters of ancient Lake Agassiz between 9,000 and 13,000 years ago. Sand and silt were carried to the lake by water from the melting glacier, then heaped up into beach ridges by waves and ice. The shoreline moved back and forth over centuries as the climate changed, causing lake levels to rise and fall. Ten tallgrass prairie areas larger than 2,000 acres remain of the 609,000-acre Agassiz Beach Ridges. About 75,000 acres of prairie areas exist, most in small parcels. Along the ridges plants and grasses are found which are suited to dry prairies, while at lower levels rare fens and marshes occur, transected occasionally by gorges lined with trees. The former prairie landscape is severely broken up by fields, homesites, powerlines, highways, and gravel quarries. An inventory of a single site—Bluestem Prairie in Clay County—revealed more than 300 different types of native plants, including pasque flowers, spiderworts, blazing star, coneflowers, purple prairie clover, and many types of sunflowers. The greater prairie chicken finds its most productive habitat in this landscape.

(Photo by Diane Wade)

Old reaper used to harvest grain. (Photo by Loren Oslie)

I was born here. The landscape is beautiful, and I love raising livestock. I just couldn't imagine raising my boys anywhere else.

Larry Woodbury, Sheyenne Delta rancher

Chapter 2

Thinking Like a Prairie
Living Laboratories

This region is truly a food basket of the world, with just priceless soils. And where did those soils come from? They came from the prairie. It would be a terrible shame to lose the ecosystem that provided us with those soils. I also think of the prairie as a gene bank, in which the genetic material from native species might have importance for developing new crop plants and medicines and things of that nature.

Gerry Van Amburg, prairie ecologist, Concordia College, Moorhead, Minnesota

More than a century after the first settlers dug their spades in Red River earth, the Red River Valley continues to deliver on its promise of opportunity. The agricultural productivity of its prairie soils has transformed the region into a land of superlatives: North America's breadbasket, its Fertile Crescent, a cornucopia of plenty.

Increasingly, however, people are looking to the margins of the black-soil valley floor, to the wind-bitten beach ridges and sandy

Prairie blazing star (*Liatris pycnostachya*) decorates an unplowed prairie in the Red River Valley, Traverse County, Minnesota. (Photo by Keith Wendt)

deltas, for another kind of promised land—some of the continent's finest expanses of native tallgrass prairie and parkland. Here, where thin, stony soils and networks of wetlands made plowing difficult, native grasslands were preserved as hunting grounds or pastures and meadows for livestock and hay. Our renewed appreciation for these leftover lands comes at a time of growing awareness of their economic, ecological and cultural importance.

A thorough study of Nature's crops and Nature's ways of making the most of a sometimes adverse environment is of scientific importance. . . . It furnishes a basis for measuring the degree of departure of cultural environments from the one approved by Nature as best adapted to the climate and soil. How unlike the more delicate, annual crops of man. . . . Nature's crops are adjusted to fit into periods of dry cycles as well as wetter ones. These have recurred again and again throughout the centuries.

John Weaver, prairie ecologist,
North American Prairie, p. 195

Almost half the region's native grasslands are set aside on public lands for wildlife, ranching, and recreation. Private landowners retain substantial stretches of prairie and parkland. Nature preserves—public and private—contain a tiny fraction of the rest. Together, these lands are vital to the long-term health of the region's native plants and animals. They provide important buffers, connecting corridors for plants and wildlife, and irreplaceable core habitat.

The economic potential of conserving and thoughtfully using these remnant grasslands is just beginning to be understood. Changes in our culture, our economy, and our understanding of ecology have made us regard prairies in a new light. Leading the way are prairie residents and communities. From rural hamlets to university laboratories, people from all walks of life are demonstrating how the sustainable uses of native grasslands can contribute to a viable economy without depleting the resources that make the Red River region a place of beauty and abundance. The traditional grassland and parkland economy—livestock and hay production, outdoor recreation, education, research, and, in recent years, a variety of small niche businesses—is being further revitalized and diversified by entrepreneurs rethinking the use of regional prairie resources.

Common to all of these enterprises is the belief that thinking like a prairie—conserving resources, finding strength in diversity, and protecting and replenishing the sources of life—makes for stronger economies as well as healthier ecologies.

Science faculty and students from several local colleges and universities find living laboratories in the tallgrass prairie and parkland. Research programs sponsored by universities add to the sci-

16

entific knowledge of these complex habitats while attracting dollars from outside the region to support the work.

The region's institutions most active in prairie and parkland research are the University of Minnesota-Crookston, Moorhead State University, Concordia College-Moorhead, North Dakota State University at Fargo, the University of Manitoba, and the Manitoba Museum of Man and Nature. Funding for research is derived from federal, state and provincial agencies, foundations, nonprofit organizations, and corporate sponsors.

Increasingly, research on native prairie focuses on practical questions, from investigating grazing methods that groom and revitalize native grasslands, to developing ways of managing soils, waters, and land to increase their health and wildlife, to discovering methods to discourage pests and noxious weeds.

A Dakota skipper (*Hesperia dacotae*) rests on a wood-lily (*Lilium philadelphicum*). This rare butterfly is found only where unplowed prairie sustains the grasses its larvae feed on, and where adults can find wildflowers supplying nectar. (Photo by Robert Dana, courtesy of State of Minnesota, Natural Heritage and Nongame Research Program)

The Value of Research

Despite evidence to the contrary, a popular misconception persists that all grasshoppers are detrimental to food crops. In 1989, scientists working for the University of Minnesota Extension Service and The Nature Conservancy found that of the more than thirty species surveyed on Bluestem Prairie, a large Conservancy preserve near Moorhead, Minnesota, just four species were considered agricultural pests. In fact, the other twenty-six species of grasshoppers never even leave the confines of the native prairie. This knowledge enables pesticide users to avoid broadcast poisoning which can wipe out the tremendous variety of prairie insects and harm birds that eat insects and to focus instead on the specific grasshopper pests where they are found.

 Researchers asking questions about groundwater, the behavior of animals, and the agricultural or medicinal value of prairie plants are searching for the answers in the intriguing territory of prairies and parkland.

Field researchers install groundwater monitoring wells at Pembina Trail Preserve Scientific and Natural Area. (Photo by Richard Johnson)

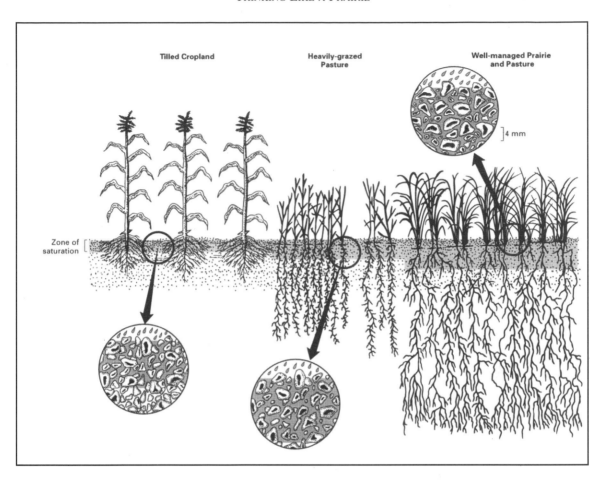

Good Earth

Researchers are finding that tallgrass prairie habitats generate and renew soil in ways that are particularly beneficial to plants. By studying these habitats, scientists are beginning to understand the role of microorganisms in improving the productivity of agricultural soils and restoring damaged forests and croplands. Working under a grant from the Northwest Area Foundation, a team of researchers, farmers, and advisors from four states, including Minnesota and North Dakota, has found important clues to the secret of healthy soils by studying the soil structure of prairies, croplands, and retired grasslands. University of Minnesota researchers Dave Huggins and Deborah Allan and consultant Margaret Jones led the effort in Minnesota.

Prairies have dense networks of underground roots. Breaking down these roots when they die are millions of microbes—bacteria, fungi, and nematodes—which create organic matter and release substances that glue the soil into chunks the size of a small nail

Rain-water penetrates quickly and deeply into the soil of well managed prairies and pastures. Relatively large clumps of particles help create and shore up tiny passageways in the soil into which water can easily flow.

19

These nuggets hold their shape even when wet. Ants, beetle larvae, and nematodes bore passageways through these clumps so that air and water can penetrate every crevice. New roots grow easily through this porous soil structure, absorbing nutrients and water that activate vigorous plant growth.

These clumps of soil particles sometimes make up almost half the prairie's soil and account for the prairie's superb ability to regulate moisture. Researchers found that the porous prairie soil absorbs and holds more water and releases it more gradually than conventionally tilled cropland. A sample core of black silt loam from tilled cropland revealed a surface soil that was more tightly packed and had smaller clumps than prairie soil. This density reduces the soil's ability to absorb water and make it available for plant growth.

Some farming practices work against creating beneficial soil structures. Fertilizers and herbicides along with certain kinds of cultivation can suppress the worms, bacteria, and fungi that build up soil structure and keep it permeable.

Planting fields to grass is an excellent first step to revitalize depleted soils. Summarizing more than a decade of research in its

Grazing by cattle, bison, and other animals is part of a prairie's natural cycle—if it encourages a variety of native plants and discourages exotic species and noxious weeds. (Photo by Don Breneman, courtesy of University of Minnesota Extension Service)

1948 *Yearbook of Agriculture: Grass*, the United States Department of Agriculture suggested: "One of the most effective ways to build up the content of organic matter in soil is to grow grass on it." The Northwest Area Foundation study proved the wisdom of this age-old advice. After examining the soil structure of lands planted to grass and enrolled in the United States Conservation Reserve Program (CRP), researchers found signs that the CRP soils had begun to revert to the beneficial clump structure of prairie earth, thanks to a doubling of microorganisms over the six to seven years since the grasses were planted.

Good Eating

A prairie's vigor and diversity–once dependent on periodic fires and grazing animals, such as bison–are an advantage to both people and prairie. Researchers in both Canada and the United States are studying ways in which rotational grazing by cattle and domesticated bison can help "groom" the prairie and mimic the natural disturbances necessary to revitalize it.

Prairie cord-grass (*Spartina pectinata*) blanketing a windswept wet prairie. (The Nature Conservancy photo archives)

Plumes of Indian grass (*Sorghastrum nutans*) with ripened seed. (The Nature Conservancy photo archives)

Seeds of the grass, needle and thread (*Stipa comata*), may plant themselves in the ground when changing humidity curls and uncurls the twisting awns. (Photo by Tom Spiekermeier)

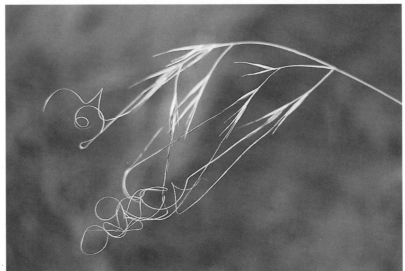

Hungry mouths also benefit from the prairie's intricate nature. In drought years, cool-season lawn grasses and pasture plants with shallow roots die unless they are watered. Prairies, on the other hand, can usually provide a reliable source of nutritious feed without costly irrigation or fertilization.

How does a prairie do it? A healthy prairie can withstand drought and resist disease—even while being grazed—due to the variety of grasses that comprise its intricate mosaic. If one grass is stricken, other species fill its place. If drought forces some grasses into dormancy, the rest grow to produce food for the animals that inhabit the land.

Making a Good Thing Better

Properly treated, damaged prairie and parkland can improve over time. Richard Pemble, professor of biology at Moorhead State University, has researched the ecology of Red River prairies and other habitats for more than two decades. Pemble recently concluded a project involving more than seventy study plots located in prairies managed by The Nature Conservancy and the Minnesota DNR's Scientific and Natural Areas Program. Managers use fire and other treatments in multiyear rotations.

In 1995, Pemble revisited the 100-square-meter plots that researchers had surveyed in 1979 and 1980. His goal: to document changes in the plant communities over the past fifteen years. The results are encouraging. "Over the fifteen years that these prairies have been managed by The Nature Conservancy and the Scientific and Natural Areas Program, the plots have improved," Pemble observes. "The variety of plant species seems to have increased in all stands, sometimes as much as fifteen to twenty percent. And these changes overall indicate an increase in good prairie species. Things are looking better today under the kind of management practiced by The Nature Conservancy and the Scientific and Natural Areas Program."

Richard Pemble (Photo by Michael Burian)

Wood-lily (*Lilium philadelphicum*) at Goose Lake Wildlife Management Area. (Photo by Fred Harris, courtesy of State of Minnesota, Natural Heritage and Nongame Wildlife Program)

Canada goldenrod (*Solidago canadensis*), seen at right, is one of several medicinal plants growing in prairies and parkland. Below, the double beauty of the purple cone-flower (*Echinacea angustifolia*) lies both in its appearance and usefulness to people. (Photo by Carmen Converse, courtesy of the State of Minnesota Natural Heritage and Nongame Research Program)

Sandhills pasture with penstemon (*Penstemon grandiflorus*) in the Sheyenne National Grasslands. (Photo by John Challey)

Prairies have evolved over thousands of years to find strength in diversity. In his book, *North American Prairie*, John Weaver, plant ecologist at the University of Nebraska, points out that "the possibility of so many plants growing in a given area, often 200 to 250 individuals in a single square yard, is due largely to the fact that the roots absorb at different soil levels, the tops making their development at different heights and at different seasons of the year." The variety of plants helps, too. For example, a quarter section of prairie–160 acres–contains a mixture of thirty or more grasses and sedges. The same acreage on a nearby farm or city park is likely to have no more than a few types of plants for ground cover.

David Tilman, a University of Minnesota professor of ecology, studies competition and survival among prairie plants. During the drought of 1987-1988, he found that test plots with the greatest variety of plants lost the least biomass and returned most quickly to pre-drought production levels. "The more species of plants you have, the better your chances that some are drought resistant and can continue growing," he points out. "Prairie plants respond to opportunity. Some plants conserve their roots by going dormant and are able to recover quickly after drought. After the first good rain, they grow as if nothing happened."

Good Medicine

The more diverse an economy, the more opportunities people have to start new businesses and fill jobs that match their skills, interests, and lifestyles. Such opportunities make communities interesting and stable places in which to live and work.

A prairie has its own economic diversity and offers some opportunities yet to be discovered, among them healing medicines. Roughly forty percent of the medicines used today are derived from plants and animals. Only recently have pharmacists and chemists begun to investigate the great body of medical knowledge amassed by prairie Indians to relieve pain or fight infections. The hundreds of plants, the thousands of insects, and the various birds, mammals, and other prairie wildlife might be potential sources for remedies or health enhancers.

Gerald Fauske, an entomologist at North Dakota State University in Fargo, studies prairie insects and plants. He believes native plants are valuable not just for the substances they contain but for the processes by which they make certain chemicals, what he calls their "cellular machinery." Unlike the natural substance found in willow bark on which aspirin is based, most plant remedies are poorly understood and difficult to replicate in the laboratory. Fauske points to the Pacific yew tree, the source of a chemical that is effective in treating some forms of cancer. "They can now synthesize that chemical in the lab," he says, "but researchers would never have known about its existence or discovered how to manufacture it until they found the plant in nature that produced it."

Small white lady's-slipper (*Cypripedium candidum*), a plant of wet prairies and calcareous meadows. (Photo by Bob Steinauer)

Prairies, Nitrogen and Climate Change

For the past fifteen years, ecologists David Wedin of the University of Toronto and David Tilman of the University of Minnesota have studied the impact of nitrogen on plant biodiversity in three abandoned farm fields in western Minnesota. They published some of the results of their work in the prestigious journal *Science*.

Scientists theorized that excess nitrogen in the environment that is carried in fertilizer dust from agricultural fields or emitted by cars and smokestacks could stimulate plant growth and, as a result, help absorb excess carbon dioxide, a gas that traps heat in the atmosphere, contributing to global warming. Wedin and Tilman have demonstrated that, on a prairie, nitrogen does promote plant growth the wrong kind. Excess nitrogen stimulates the growth of weeds at the expense of native species. In time, foreign plants such as bluegrass, reed canary grass, and smooth brome crowd out the prairie plants.

Furthermore, the nitrogen-guzzling weeds absorb far less carbon than native plants, canceling out any potential benefits for reducing carbon dioxide in the atmosphere. In other words, prairie plants deliver a bigger carbon bang for each nitrogen buck and they deliver it more efficiently. Native plants do not require great inputs of nitrogen fertilizers. The amount of nitrogen supplied naturally by lightning and prairie legumes is as much as a prairie habitat can use to stay healthy.

The researchers discovered an additional environmental downside to a weedy grassland of nitrogen-rich plants. Microbes in the soil break down nitrogen into soluble nitrates. Plant tissues rich in nitrogen, therefore, release higher doses of nitrates into the soil. These compounds wash into streams, causing algal blooms, or migrate into drinking water supplies, where they pose a health risk to infants.

At least three plants native to the Red River prairies and tallgrass aspen parkland are already natural sources for medicines used in treating certain medical conditions. Seneca snakeroot (*Polygala senega*) produces a substance that relieves bronchial congestion due to colds and flu. Some species of goldenrod, such as *Solidago canadensis*, produce an ingredient used to treat urinary tract conditions. From purple coneflower (*Echinacea angustifolia* and its relative *E. purpurea*) comes echinacea, a product readily available in health food stores in pill form. Believed to prevent common colds and generally boost the immune system, echinacea is widely used in Europe for treating respiratory and urinary infections and to speed the healing of surface wounds.

Ammonia fertilizer is applied to cropland in the spring. (Photo by Don Breneman, courtesy of University of Minnesota Extension Service)

Chapter 3

Thinking Like a Prairie
Natural Services and a Conservation Economy

On soil without a cover of vegetation or debris, the impact of raindrops . . . on the surface loosens the soil particles, which are then suspended in the water. This water on entering the soil pores carries the fine, suspended particles with it. The pores are more or less completely clogged on the surface, and a compacted layer of soil formed. This layer greatly decreases absorption and increases water loss by runoff.

John Weaver, prairie ecologist,
North American Prairie, 1954

Enough Clean Water

Having a dependable, abundant source of clean water is critical to the Red River region. New agricultural processing plants, for example, which rely on good water supplies, are boosting the region's economy and promising a greater local return on commodity investments. Since precipitation amounts to less than twenty-four inches a year in this part of the country, practicing maximum conservation and protecting land that absorbs and stores groundwater is common sense. Prairie, parkland, and wetlands–an integral part of the region's natural hydrology–are effective aids for protecting its water supply.

A researcher studies the diversity of prairie plants in the Agassiz Beach Ridges. (Photo by Keith Wendt)

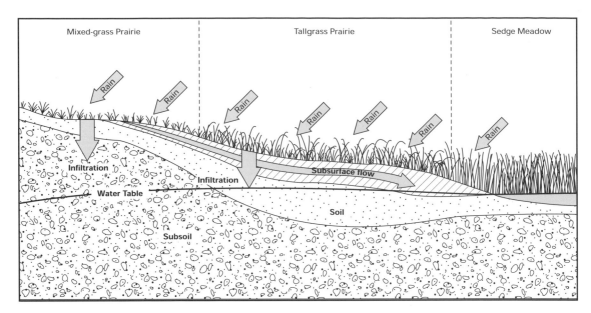

Mixed-grass Prairie | Tallgrass Prairie | Sedge Meadow

Much of the rainfall in the Sandhills of the Sheyenne Delta reaches the water table through the level areas where the tallgrass prairie grows. On the ridges and steeper slopes, less of the rainfall becomes groundwater.

Soil and water researchers David Hopkins (left) and Jimmie Richardson. (Photo by Michael Burian)

North Dakota State University researcher David Hopkins spent five years studying the water-storing capacity of prairie and wetlands by measuring changes in the groundwater levels of the Sheyenne Delta. The prairie habitats play an important role in stabilizing the landscape and keeping the aquifer pure. The delta landscape consists of upland mixed-grass prairies and low-lying tallgrass prairies and sedge meadows. Hopkins observed that a large amount of the rain and snowmelt entering the Sheyenne Delta aquifer probably does so through the large, extensive flat areas where tallgrass prairie typically grows.

Winter scene along the Sheyenne River bottoms. Groundwater seeping into the bottoms creates fen habitats and other unique environments for rare plants. (Photo by Loren Oslie)

Extreme droughts and major floods in the Red River region during the twentieth century (below). (Sources: Krenz & Leitch 1993; U.S. Army Corps of Engineers and Minn. Dept. of Natural Resources 1995; Swerman, Baker and Skaggs 1987)

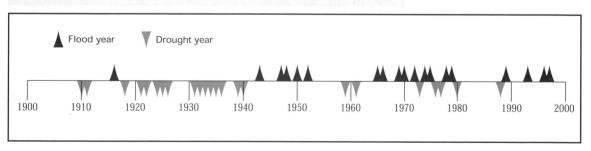

Prairie cord-grass and other plants stabilize the banks of a stream. (The Nature Conservancy photo archives)

His colleague Jimmie Richardson, soil science professor, has observed the same thing wherever prairie is found. "The prairie stores nitrogen and carbon and increases soil structure," he says, and that makes for better water absorption. The prairie takes in water and slowly releases it–making the ancient shoreline that rings the Red River Valley a great recharge zone. Speaking of the prairies on beach ridges in Minnesota, Richardson said, "There's no runoff, except in the spring when the ground is frozen. Water flowing across the ridges sinks into the water table. If you have good aggregates [soil clumps], soils can absorb water, and it will move laterally through the soil."

Good aggregates depend on grass roots and carbonates–akin to Alka-Seltzer–that form in the soil. As plants take up water, they dry the soil, unleashing a chemical reaction that deposits calcium carbonate in the soil. This mineral locks organic matter in the soil and gives it a dark color–all of which means better soil structure, better absorption of water, and better filtering of water. Says Hopkins, "The Sheyenne Delta sits on the largest unconfined aquifer in the eastern Dakotas. The water is very pure."

Since 1900 severe drought has visited the Red River region in twenty-three out of ninety-six summers. Many communities in the Red River region depend on groundwater for drinking and other uses. Finding ways to increase water-absorbing prairie acres could help preserve aquifers in the Red River Valley.

Slowing the Floods

While nothing could have stopped the terrible flood of spring 1997, periodic flooding has been worsened by development, the plowing of prairies, and the drainage of wetlands that once served as natural containment areas for excess water. One expert calculated that Fargo-Moorhead, for example, weathered thirty-one major floods between 1897 and 1997. Of these, eighteen occurred after 1965, and eight after 1985, a period that included severe drought from 1987 to 1989.

Prairie ecologist Gerry Van Amburg, who serves on the Buffalo-Red River Watershed Board, draws a connection between this flooding and the loss of prairie wetlands:

> I looked at this area from a plane during the 1993 flood, and I can honestly say that if we hadn't had all the wetlands we have out there—even though they are half of what used to exist—there

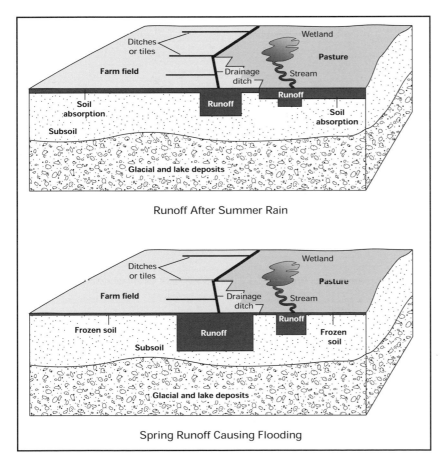

Runoff After Summer Rain

Spring Runoff Causing Flooding

After summer rains, prairies and wetlands hold back water and release it slowly. Tiles and ditches carry water away more quickly. In early spring, water streams off the frozen landscape, causing flooding.

would have been a whole lot more water pouring down into Fargo and Moorhead. That flood would have been even worse. We've changed the landscape to such an extent that it certainly has aggravated the problem.

The argument that engineers use is that "once a wetland is full, it's full." And if it rains, the water will just run over and come down anyway. Well, that's true, but most wetlands we have around here tend to draw down at the end of the summer. When spring comes, these prairie potholes do have some capacity to store water. But since we've wiped out more than half of them, the water is forced to run down off the ridge and into the valley, where it floods farms. Ultimately, it enters the streams, carrying with it a lot of nutrients and sediment. Wetlands hold back sediments and trap nutrients so that they don't run off into streams and cause problems. Even if we had all our original wetlands, we wouldn't eliminate flooding completely, since it's also driven by climate. But I think many people have realized that we've drained a lot more than we should have.

Unlike prairies, tilled and ditched croplands release water quickly following a snowmelt or rainstorm, and cities and towns—with their many impervious rooftops, roads, sidewalks, and parking lots—shunt water away even faster. The result is too much water in wet weather, and no water to spare during dry spells or even in a normal August—many creeks and rivers simply run dry.

Prairie and parkland, bolstered by wetlands, can help reduce flooding during wet years and even out the flow from month to month during normal years. With their porous soils, prairies and well-managed pastures act as giant sponges, soaking up water and filtering it before it percolates into the groundwater. The leaves and matted dead plants detain water, giving it more time to evaporate. Plants pull water up from their roots and release it through their leaves, a process called evapotranspiration.

As a model for integrating human and natural systems, Red River communities can look to places such as Prairie Crossing, located an hour's drive from the Chicago Loop. This new development has made national headlines for its novel approach to subdivision planning. Only one-fifth of its 667 acres is used for housing; the remaining land is devoted to farmland, a restored prairie, organic gardens, and community space. According to the *Wall Street Journal*, Prairie Crossing and other developments like it across the country reflect "the growing disillusionment with suburban sprawl and a willingness to pay a premium to live in a place that protects the land."

Prairie Crossing is acknowledging what cities with a well-developed network of green space have known for a long time—that open space is good for the soul as well as the economy. Stable and

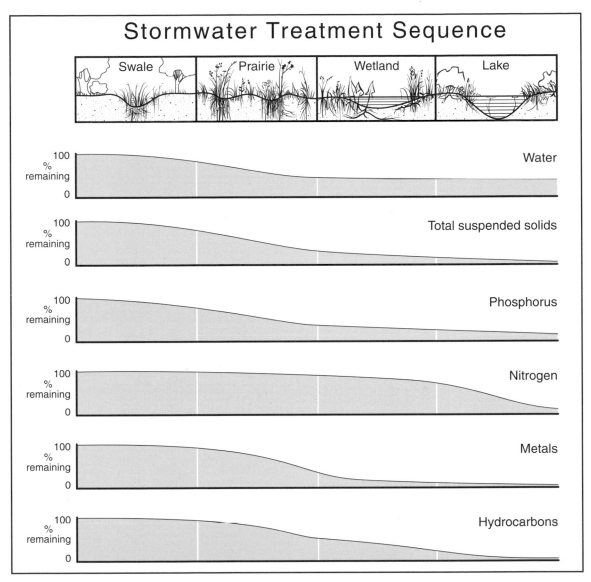

Stormwater Treatment Sequence

| Swale | Prairie | Wetland | Lake |

Water — 100 % remaining — 0

Total suspended solids — 100 % remaining — 0

Phosphorus — 100 % remaining — 0

Nitrogen — 100 % remaining — 0

Metals — 100 % remaining — 0

Hydrocarbons — 100 % remaining — 0

enhanced property values pay long-term dividends on a community's investment in public open space.

New communities like Prairie Crossing are building public value in other ways as well. Their green space serves as natural infrastructure as well as a recreation area, a human food source, and wildlife habitat. Instead of whisking stormwater from the site directly into rivers and streams via a costly system of culverts and sewers, Prairie Crossing will use a series of created prairies and wetlands in a four-stage runoff-filtration system. This green engineering slows the course of stormwater, allowing sediments and pollutants to be filtered near their source. The dense growth of

Water running off parking lots, buildings, and streets can be cleaned of most nutrients and pollutants by passing through grassy swales, prairies and wetlands that are planted as part of a new development or re-building program. (Values are estimates. Illustration by Rob Dunlavey, modified with permission from Steve Apfelbaum)

prairie grasses and other plants filter up to ninety percent of sediments and seventy percent of the nitrogen carried by stormwater runoff. A patch of prairie also can absorb up to three-quarters of the phosphorus in runoff, the substance that causes algae to bloom and gives water a bad taste.

Steve Apfelbaum, an ecologist with Applied Ecological Services in Brodhead, Wisconsin, and designer of Prairie Crossing's natural filtration system, says, "In our models based on scientific literature, we demonstrate the capacity prairies have for storing and cleaning dirty water. We estimate that grasslands can remove up to sixty-five percent of the urban runoff. It's nice to tell a developer or city planner they can save money on sewers and beautify their projects by planting prairie."

Prairie and wetland restoration can be especially beneficial to the Red River region. Increased frequency and severity of flooding in the Red River watershed is a growing concern of the people—and their pocketbooks—in this region. To mitigate the severity of flooding, the region's watershed management groups propose, among other solutions, building thirty-three new impoundments on tributaries of the Red River. Constructing and repairing these dams will be costly. Even with this investment in engineered infrastructure, downstream residents on the Red River will experience virtually no decrease in 100-year flood peaks. Moreover, an environmental impact statement completed in 1996 says that fish and aquatic life in the Red River basin could be harmed if all the dams are constructed.

Lake sturgeon spawning in a river tributary. Free-flowing streams allow fish such as the sturgeon and dozens of others to complete their life cycle in the Red River and its tributaries. (Photo by Shawn Johnson, courtesy of State of Minnesota, Department of Natural Resources)

The Red River flooded even before the prairies were plowed and wetlands drained. Flooding, by itself and within limits, is not harmful to the land—on the contrary, it is often beneficial—except when uncontrolled and as massive as recent flooding. Many benefits of these systems can be restored by placing more land in enrollments such as the Conservation Reserve Program, by planting prairie in the sandy and gravelly beaches and deltas that serve as groundwater recharge zones, by restoring drained wetlands, by incorporating prairies into housing and commercial developments, by planting grass-lined waterways in fields, and by installing grass filter strips at streamsides. An added benefit is a boost to regional fish and wildlife. These approaches also expand options in any flood control plan for the Red River Valley.

An Ancient Tradition

Hunting, one of the oldest human activities in the Red River region, is still an important part of the culture and economy. Communities located within the Tallgrass Aspen Parkland, in particular, have long recognized the value of wildlife to local and regional economies. Many residents consider wildlife central to their quality of life. As Kittson County Commissioner Joseph Bouvette, a dentist and lifelong Kittson County resident, observes, "If you like the outdoors, it's great up here. I love to hunt and fish. It's nice to be able to get off work and be out hunting within half an hour. People from the Twin Cities have to wait until Friday, drive six hours to get here and then drive back."

Despite the distance, Bouvette points out that the area has drawn consortiums of hunters who have purchased tracts of tallgrass aspen parkland for private hunting preserves. The availability of hunting grounds adjacent to the region's numerous state wildlife management areas has made these lands especially desirable. In the mid-1980s, for example, when the numbers of moose peaked at three to four animals per square mile, the Minnesota Department of Natural Resources (DNR) issued nearly 500 moose-hunting licenses annually.

Wildlife is an important part of the life and economy of communities across the border as well. The agricultural regions of southern Manitoba host the greatest diversity of migratory game bird species in the province. According to the Manitoba Department

Kittson County Commissioner and hunter Joe Bouvette. (Photo by Michael Burian)

Beaver and other fur-bearers provide income for trappers in the Red River region. (Photo by Bill Berg, courtesy of State of Minnesota, Fish and Wildlife Division)

of Natural Resources, nearly forty percent of Manitoba's annual fur harvest comes from the 8,500 trappers registered in southern Manitoba. The department also reports that hunting remains a significant generator of revenues. About ten percent of the adults in Manitoba are licensed hunters, spending an annual average of fifteen days hunting, as well as an average of $850 on related products and services.

The Growing Tourism Enterprise

Like the people who endure long, harsh winters, the prairies and parkland pack a lot of activity into their short warm season. This burst of energy is visible in the almost daily changes of color, texture, sights, and sounds that unfold from early May through September over the grassland. Prairie visitors are often reminded of the saying that no one steps into the same river twice; the prairie changes constantly, too.

Native prairies are some of the best-kept secrets of the tourism industry, but word is spreading. People are traveling great distances to quench a longing for the beauty and solitude of open spaces, where they have a chance to see certain plants, mammals, birds, and butterflies perhaps only once in a lifetime.

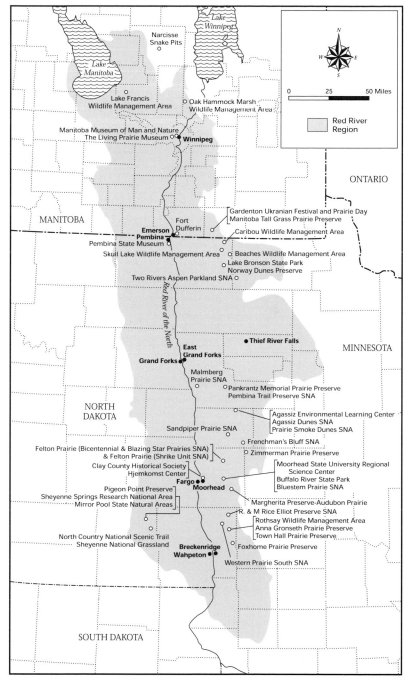

Tourism destinations in the
Red River region.

In Manitoba, the Department of Natural Resources is beginning
to recognize the economic potential of people who want to experience wildlife through the lens of a camera rather than the barrel of
a gun. More than eighty-five percent of Manitobans engage in
wildlife-related activities, spending more than $150 million on

such experiences as nature outings and bird feeding. The department estimates that ecotourism could contribute more than thirteen million dollars to provincial coffers.

Some of Manitoba's most successful ecotourism efforts are located in the Tallgrass Aspen Parkland. The Narcisse Snake Dens, north of Winnipeg, for example, draw more than 13,000 people each spring to witness thousands of red-sided garter snakes in their annual courtship ritual. More than 70,000 visitors each year tour the nearby Oak Hammock Marsh Wildlife Management Area, lured by the opportunities to view prairie wildflowers and marsh birds. In 1989 the Manitoba Department of Natural Resources and Ducks Unlimited Canada began a collaboration to build an eleven-million-dollar wetland conservation center in the Oak Hammock Marsh preserve. The center houses interpretive facilities, as well as offices for more than 100 Ducks Unlimited staff.

Not all ecotourism opportunities are located in outlying prairie remnants. The Living Prairie Museum, a 12.8-hectare reserve in Winnipeg, offers a green respite in an urbanized landscape dotted with roads, schools, and an industrial park. The museum's education coordinator, Jonina Ewart, points out that people come from all

Children learn about prairie plants and animals at The Living Prairie Museum in Winnipeg. (Photo by Kristina Hurtubise, courtesy of Living Prairie Museum)

Unlocking the Door to Your Own Backyard

As economic development director for five Polk County cities, Wayne Goeken helped establish the Agassiz Environmental Learning Center (AELC), which has created local jobs and drawn visitors to Polk County. The center introduces the magnificent natural features of the area to casual visitors as well as to scheduled classes and groups. Nature trails wind through pristine oak savannas, sand dunes and along the Sandhill River, which remains open even in severe winters. A historic Red River oxcart trail, which once ran down the main street of Fertile, traverses the area. In the mid-1800s, the Red River trails provided links between Winnipeg, Manitoba, and Pembina, North Dakota, and St. Paul trading posts. The Nature Conservancy's Agassiz Dunes and the state of Minnesota's Prairie Smoke Dunes Scientific and Natural Area lie just to the south.

The AELC receives its funding from five cities, the county, the local watershed district, the Northwestern Minnesota Initiative Fund, and other sources. In 1993, the AELC built a nature center with the help of local volunteers, financed with a state outdoor recreation grant and matching dollars from the city of Fertile. The AELC averages 2,000 to 3,000 visitors a year and has 120 members. Volunteers and members provide programs for the public and surrounding school districts. The 1996 Minnesota State Legislature approved $300,000 for the design of a residential facility to accommodate overnight groups.

Sketch of an oxcart train passing by Glyndon, Minnesota, around 1860 by pioneer settler Orabel Thortvedt. (Courtesy of the Minnesota Historical Society)

over the world to observe the prairie's wildflowers, birds, and other species. The Manitoba Museum of Man and Nature in downtown Winnipeg offers visitors an opportunity to learn more about the prairie with spectacular natural history dioramas that focus on the province's native landscapes, including the prairie and the tallgrass aspen parkland ecosystems.

In the Red River region, history buffs and nature lovers now can find a round-robin of destinations that provide a glimpse of what life on the prairie was like more than a century ago. An exciting new tourism circuit is developing along a route that extends from southern Manitoba along the Agassiz Beach Ridges and into southeastern North Dakota. Coordinated marketing efforts among cities and agencies of the region are assisting nature travelers and local businesses at the same time.

Gardenton, Tolstoi, and Stuartburn, Manitoba, form the nucleus of a tourism hub that promotes the cultural and natural history of the Tallgrass Aspen Parkland region. The town of Emerson, Manitoba, twenty-five miles southwest of Gardenton and three miles north of Pembina, North Dakota, has plans to restore Fort Dufferin, the site for several annual heritage events and recreational trails. Similarly, the Hjemkomst Center and Clay County Historical Society at Moorhead,

Metis Indians and frontiersmen relax on the trail by their Red River carts around 1860. The oxen or horses were turned out to pasture where the men stopped for lunch or camped for the night. The 400-mile journey from St. Paul to Pembina took several weeks, but between 1840 and 1870, it was the most economical way to move buffalo hides and honey south, and bring store goods back to the settlements. (Photo by Upton, courtesy of the Minnesota Historical Society)

Minnesota, celebrate the Norwegian heritage of the area and its major tallgrass preserves. And Pembina, North Dakota, recently opened a new $2.2-million state museum. Visitors traveling between these places can easily find native prairies that look much like they did when the first settlers arrived. New hiking and biking trails, linking geologic, cultural, and historic points of interests, also are fueling a growing segment of the tourism economy. They also offer exciting new opportunities for local residents to enjoy the outdoors.

Many local governments are finding economic advantage in educational programs on the geologic formations and rare habitats in their jurisdictions. In Kittson County, Minnesota, for example, the local Soil and Water Conservation District purchased eighty acres of tax-forfeited land in 1995 to use as an outdoor classroom for students in a three-county area. And the Agassiz Environmental Learning Center in Fertile, Minnesota, serves local students as well as schoolchildren from around the state.

Visitors to Loewen Prairie near Gardenton, Manitoba, enjoy a day on the prairie. (Photo by Marilyn Latta, courtesy Manitoba Naturalists Society)

Growing Prairie Seeds and Plants

Native plants have evolved over thousands of years to local soil, light, and climate conditions. As a result, native plants are hardy and require almost no maintenance. Companies who specialize in native-seeds occupy a niche market whose customers include landscape designers, prairie restoration companies and organizations, home gardeners, transportation agencies responsible for roadside beautification, and enrollees in the Conservation Reserve Program who want to plant a cover crop resembling native grassland.

Homegrown Prairies

Since 1991, The Nature Conservancy has been restoring unproductive cropland in the Agassiz Beach Ridges to prairie. As of 1996, the Conservancy had converted 262 acres to prairie and has plans to add another 1,276 acres by 2002. These restored grasslands serve as protective buffers in core preserves. Whenever possible, the Conservancy leaves the surrounding land undeveloped or encourages complementary and economically viable uses—including businesses such as Bluestem Farm.

Brian Winter. (Photo by Michael Burian)

Purchased from The Nature Conservancy in 1995 by Prairie Restorations, a prairie seed grower and restoration company in Princeton, Minnesota, Bluestem Farm is a 500-acre tract adjoining the Bluestem Prairie. The operation will redevelop former cropland as fields for the production of prairie seeds.

Brian Winter, science and stewardship director for the Conservancy in the region, says, "Businesses like Bluestem Farm will help fill a need for high-quality seed in the region, while protecting the Conservancy's core prairie preserves." Conservancy scientists also benefit from the opportunity to experiment with a variety of restoration techniques.

Winter explains that current restoration procedures begin with burning remnant prairies, a revitalizing force that causes plants to flower. Seeds are then harvested with a combine. The gathered seed is either broadcast on the fields and then worked into the soil or stored over winter and planted in the spring. These seeded acres are mowed annually for two years. After enough dead and decaying vegetation has built up on the ground, prairie managers renew plant growth with controlled burns. "It seems to be working," says Winter. In some of the Conservancy's oldest prairie conversions, researchers have tallied up to seventy different species of plants.

A related business, the herb industry, is growing at fifteen to twenty percent per year, according to a leading United States business journal, and prices for the dried root of the purple coneflower (a tallgrass prairie plant) are around fifteen dollars a pound with the extract retailing at twelve to twenty-six dollars an ounce. The Sahnish Farms Cooperative, run by an American Indian group in New Town, North Dakota, is ready to enter this market with medicinal plants for the homeopathic industry. Sahnish Farms' president, Robert Lattergrass, a marketing management instructor at Fort Berthold Community College and former economic development director of the Turtle Mountain Chippewa Tribe, says that, in the venture, they hope to become processors of botanical drugs.

Bison Are Back

The historical custodians of prairie were bison and the people who hunted them, sometimes with fire. Bison grazing habits produced a mosaic in the vegetation that benefited a wide variety of grasses, birds, butterflies, and other species, while fire rejuvenated the bison's forage.

People are again using bison, and they are finding many advantages. For example, bison are better adapted to grazing the prairie than cattle. About ninety-five percent of a bison's diet is grass, com-

pared to eighty to eighty-five percent for cattle. To find enough fresh grass to support the herd, bison have evolved to move almost continuously over the land. Since prairies are about ninety percent grass, bison are much less discriminating about where they put their mouths. They simply grab a mouthful of grass and move on. Cattle, on the other hand, tend to look for and quickly eat plants such as beans, peas, lilies, and other forbs, thus leaving less desirable plants to flourish, changing the plant life of the prairie.

Bison are also well adapted to the climatic extremes of the northern plains. They slow their metabolism to conserve energy during the winter, for example, and withstand summer heat without shade better than

Bison (top) pause atop a ridgeline before moving on to graze. (Photo by Harold Malde) Bottom: Cattle have the ability to look for and eat the plants they prefer. (Photo by Jeff Printz, courtesy of Natural Resources Conservation Service)

45

In large pastures, bison tend to eat more grass than cattle because of their habit of grazing on the move.

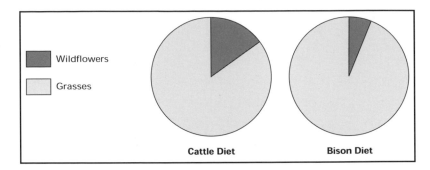

cattle. Andy Schollett, former bison manager on the Ordway Prairie in South Dakota, recalls some of his first experiences with bison: "The temperature reached 108 degrees, and the bison continued to do what they do when it's sixty degrees: they grazed, they wallowed. My first winter at Ordway, I went out and chopped holes in the potholes for water, but those bison were licking the snow and didn't go near the water, so I abandoned the idea of keeping the water open. They're basically a low-maintenance animal."

"On the northern plains, bison have tremendous economic advantages over cattle because they can withstand the winters, and bison don't require expensive shelters that have to be built for cattle," says Dennis Sexhus, chief executive of the North American Bison Cooperative at New Rockford, North Dakota. Sexhus sees a growing demand for bison products coming from customers throughout Canada, the United States, and Europe. The cooperative expects to handle up to 10,000 animals in the next year or two, twice the number it processed in 1995.

Still other private landowners keep bison because of the romantic link to a colorful past, a time when herds of these immense animals traversed the Red River Valley in search of greener pastures. Bison are a successful chapter in conservation, brought back from the brink of extinction, and now multiplying under the care of people who value their unique adaptations to local conditions.

Grazing Native Prairies

Prairies multiply their opportunities for survival by growing grasses in two intervals: during the spring (cool-season grasses) and during the summer (warm-season grasses). Spring grasses grow quickly in May and June and make seed in the summer. These grasses include blue-joint, wheat-grasses, wild-rye, June grass, and several kinds of needle grasses or porcupine grasses. Many grasslike plants, known as sedges, also are cool-season.

Summer grasses flourish in hot weather and make flowers and seeds in late summer and fall. These warm-season grasses include

Knowing a Good Thing When You See It

When diversity in a pasture improves, a grazer can see the difference. A team of twenty-four people including farmers, researchers, consultants, government agencies, and non-profit farm-support organizations is studying diversity on pasture and crop fields in a project funded by the Minnesota Institute for Sustainable Agriculture and coordinated by the Land Stewardship Project of Minnesota. The team hopes to show how farm and ranch managers can put their powers of observation to work for the good of their land. By comparing intensely grazed, short-term pastures with cornfields and pastures grazed all season long, the investigators are looking for signs that a farmer or rancher can use to identify healthy land. They have found that pastures intensely grazed for short periods with rest rather than for the entire summer have more grassland birds, cleaner streams with more fish, and between fifty and 150 percent more earthworms in the soil.

big bluestem, little bluestem, switchgrass, Indian grass, prairie dropseed, cord grass, muhly grass, and sideoats grama. Since different plants in the prairie grow at different times, a rancher can move cattle into those areas that are most nutritious.

Some cattle growers take advantage of this growth cycle to lower their costs of production. In fall and spring the cool-season grasses provide good forage, and in midsummer the warm-season grasses take

Purple prairie clover (above) (*Petalostemum purpureum*), a legume of the prairie, decreases with heavy grazing. (Photo by Carmen Converse, courtesy of State of Minnesota Natural Heritage and Nongame Program)

Colonies of flickertail or Richardson s ground-squirrels (left) (*Spermophilus richardsonii*) form in grazed prairie food for hawks, badgers, and other prairie predators. (Photo by Carroll Henderson, courtesy of the MN Nongame Program)

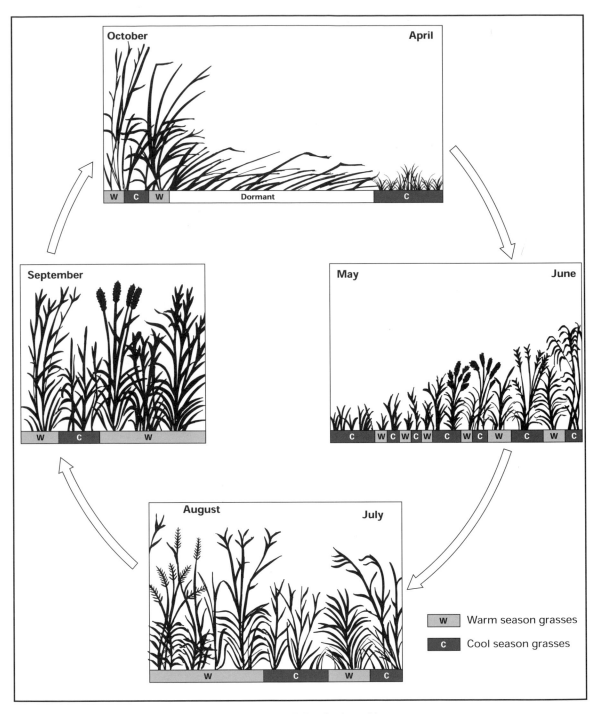

Beginning at top right: Cool spring-time temperatures bring forth prairie grasses that begin flowering in June ("cool-season grasses"), while warm summer days coincide with the rapid growth of other grasses ("warm-season grasses") having a peak blooming time of September.

over the job. At times when beef prices are low and grain prices are high, the prairie's diversity gives a farmer a little extra insurance. Or a prairie's feed value can be put in the bank by haying it. Cutting performed after mid-July can substitute for the periodic grazing and fires that naturally reinvigorated native grasslands.

Conservation

A host of conservation organizations and programs contribute to the economy of the Red River region as they work to conserve the land, soil, water, and biodiversity of the region. Foremost among these are the federal farm programs that encourage conservation through direct payments to local landowners, checks that add up to millions of dollars a year.

In the early 1980s, the United States slid into a severe agricultural crisis. To help stabilize the income of small farmers as well as address rural environmental concerns, the United States Conservation Reserve Program (CRP) was inaugurated in 1985. Under the program, farmers were allowed to sign ten-year contracts to take erodible cropland out of production and plant it in grass.

According to a report by University of Minnesota professor Steve Taff, the northwest corner of Minnesota has led the state in CRP acreage. About 270,000 acres of cropland, for example, were planted to grass and alfalfa in the narrow landscapes of the Agassiz Beach Ridges and Tallgrass Aspen Parkland alone.

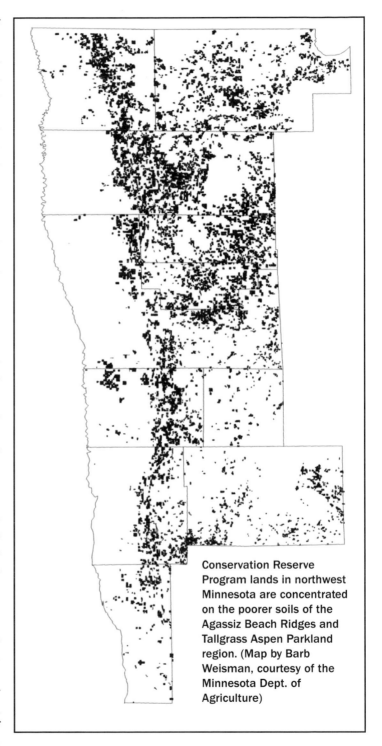

Conservation Reserve Program lands in northwest Minnesota are concentrated on the poorer soils of the Agassiz Beach Ridges and Tallgrass Aspen Parkland region. (Map by Barb Weisman, courtesy of the Minnesota Dept. of Agriculture)

49

Habitat fragmentation by roads, homesteads, cropland and a town at the Manitoba Tall Grass Prairie Preserve. (Photo by Gene Fortney)

Agricultural analysts point out that the heavy CRP sign-up in the Tallgrass Aspen Parkland is an indication of the marginal farming conditions there. This region was the agricultural frontier in the 1970s and 1980s, when thousands of acres of virgin prairie and parkland were broken for the first time, then later enrolled in CRP. For many farm operations struggling against poor drainage, a short growing season, and calcareous soils whose high pH levels limit the kinds of crops that can be grown, enrolling acreage in CRP made good economic sense. In some northwestern Minnesota counties, the CRP payment rates exceeded the rental rates for the land.

CRP has been a boon to wildlife as well as people. In ten years, the program created nearly two million acres of new grassland in Minnesota–more than twice the acreage the Minnesota DNR and U.S. Fish and Wildlife Service combined have acquired over the past forty years. In addition to providing badly needed grassland habitat for such birds as the greater prairie chicken, CRP lands have reduced erosion and helped stem the siltation of wetlands, streams, and lakes. In areas such as the Agassiz Beach Ridges and Tallgrass Aspen Parkland, CRP set-asides have also helped to heal the fragmentation of the landscape by linking wildlife habitats with grasslands that serve as green passageways. There are many other benefits not as easily seen, such as reducing greenhouse gases causing global climate change. Grasslands and brushlands store airborne carbon dioxide in their roots, and wet brushlands according to some studies take up as much as a tropical rain forest.

But the first CRP contracts expired in 1995, and almost half of the remaining contracts retired in 1997. A survey by the Soil and Water Conservation Society estimates that with cuts in the program under the new farm bill, more than half the marginal lands currently retired under CRP will be returned to production. Changes in enrollment criteria that emphasize contract cost are eliminating many CRP acres in the region. The replowing of these acres will break up the grassland landscape again and destroy badly needed habitat.

Fiscal pressures are changing agricultural policy in Manitoba as well. For example, in 1994 the Canadian Parliament repealed the Western Grain Transportation Act, or Crow Rate. Established in 1897, the rate provided government subsidies–about $700 million annually in recent years–for the shipping of selected export crops by rail. Critics charge that the policy encouraged farmers to plow every available acre, though whether or not marginal farmland will be retired or more intensively cultivated now that farmers must

Waterfowl such as blue-winged teal (below), mallards, and Canadian geese benefit from restoration of wetlands and prairies in the Red River region. (The Nature Conservancy photo archives)

bear the full costs for the transportation of their products is a matter of considerable debate.

In a time of fiscal belt-tightening on both sides of the border, there is growing recognition that conservation initiatives will depend on establishing new partnerships with private landowners on actively farmed lands. The Manitoba Habitat Heritage Corporation, an independent body established in 1986 by the Manitoba Habitat Heritage Act, works with individuals, local conservation groups, and provincial and national conservation organizations to "carry out conservation, restoration and enhancement of Manitoba fish and wildlife habitat." Among other programs, the corporation administers the North American Waterfowl Management Plan, a binational agreement to restore declining waterfowl habitats. The plan provides for the leasing or purchase of high-quality habitat, primarily in agricultural regions, while it extends incentives to private landowners for management practices that improve wildlife habitat.

The North American Waterfowl Management Plan is being supported across the border as well. In 1996 the United States Migratory Bird Commission awarded a North American Wetlands Con-

Glacial boulder at Bluestem Prairie, Clay County, Minnesota. (Photo by Steve Kuchera)

servation grant of almost one million dollars to a consortium of conservation groups, American Indian tribes, and government agencies to protect, restore and enhance prairie-wetland complexes in the Red River region of Minnesota.

Plans also call for introducing sweeping legislation in Canada that will promote the proliferation of conservation easements on privately owned land. In exchange for following good management practices that protect rare or sensitive habitats, landowners will receive government payments without relinquishing the title to their lands.

Similar initiatives have recently been introduced by the United States Fish and Wildlife Service to protect 77,000 acres of tallgrass prairie from Des Moines, Iowa, to the Canadian border. Land purchased from willing sellers would be enrolled in a national wildlife refuge system. Other acres would be added through purchasing perpetual conservation easements on private lands.

According to a report from Manitoba's Department of Natural Resources, this trend toward more public-private solutions to conservation will likely become the wave of the future:

> As wildlife continues to face new challenges to its survival, the ability of governments to respond to these challenges is diminishing. With increasing demands on the provincial treasury for health care, education, social services, and servicing the public debt, fewer staff and smaller budgets are available for wildlife management. In the future, citizens may have to participate more directly in the solutions to wildlife management problems. A growing number of First Nations and nongovernmental organizations are already pointing the way to new partnerships in wildlife management, and this trend is expected to continue.

Chapter 4

Living with the Prairie
In the Sheyenne Delta

Land is not just a surface on which a settlement is built, like a Monopoly board where little green houses and red hotels are cleverly set out. Land is the origin *of community–of the community of people every bit as much as it is the community of nature.*

Charles Little, writer, *Hope for the Land* (1992)

The Sheyenne Delta in southeastern North Dakota contains the largest blocks of tallgrass prairie in the Red River region. Formed roughly 10,000 years ago at the mouth of an ancient river carrying glacial meltwater into Glacial Lake Agassiz, it is one of the largest of the many deltas that formed at the edges of the lake. Measuring about 700 square miles, the Sheyenne is second in size only to the Assiniboine Delta in Manitoba. Characterized by dune formations shaped by the wind before native grasses stabilized the sandy but productive soils, the land is now used mostly for dryland and irrigated farming and cattle production.

Early surveyors' notebooks report that the Sheyenne Delta was a mosaic of prairie, savanna, marshes, forests, and shrub land, in which tallgrass prairie covered eighty percent of the land. For thousands of years following the glacier's last retreat, this area of eastern North Dakota was inhabited by various native peoples.

Black-eyed Susans (*Rudbeckia hirta*) adorn the low portions of tallgrass prairie in the Sheyenne National Grasslands. (Photo by John Challey)

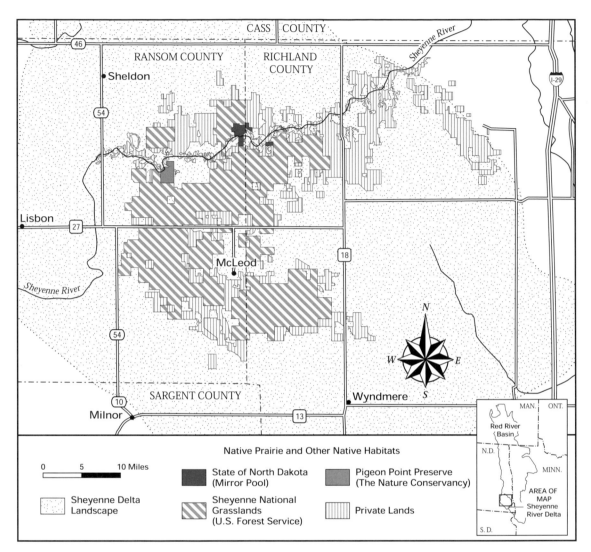

CASS COUNTY

46

RANSOM COUNTY

RICHLAND COUNTY

• Sheldon

54

Lisbon

27

McLeod

18

54

Sheyenne River

N

W E

S

SARGENT COUNTY

10

Milnor

13

Wyndmere

I-29

Sheyenne River

Native Prairie and Other Native Habitats

0   5   10 Miles

Sheyenne Delta Landscape

State of North Dakota (Mirror Pool)

Sheyenne National Grasslands (U.S. Forest Service)

Pigeon Point Preserve (The Nature Conservancy)

Private Lands

MAN. ONT.

Red River Basin

N.D.

MINN.

AREA OF MAP Sheyenne River Delta

S.D.

Native prairie in the Sheyenne Delta Landscape. (Source: The Nature Conservancy Great Plains Program and North Dakota Natural Heritage Program)

European settlement, however, brought changes to the land on a scale rivaling that of the glaciers. North Dakota was opened to farmers by the Homestead Entry Act of 1862 and by railroad companies, which sold off huge tracts of land granted them by the federal government. But one of the most formative events occurred during the 1920s and 1930s. The collapse of the nation's economy devastated delta residents. Compounding their misfortune was a prolonged drought, one in a series of cyclical dry spells common to the western edge of the tallgrass prairie.

In the 1930s, the Resettlement Administration was responsible for purchasing damaged land in the Sheyenne Delta, until it was dissolved. Later the federal government established the Sheyenne River Land Utilization Project and began managing the land under

Homesteaders breaking prairie on the Dalrymple farm twenty miles west of Fargo, 1878. (Photo by F. Jay Haynes, Fargo, courtesy of the Minnesota Historical Society)

Restoration of a sand blow-out in the Sheyenne Delta in the early 1960s. (Courtesy of U.S. Forest Service)

the 1937 Bankhead-Jones Farm Tenant Act. The Soil Conservation Service (SCS) was put in charge of these lands and given three directives: reestablish vegetative cover, change land-use practices and rebuild the rural economy. The SCS (now called the Natural Resources Conservation Service [NRCS]) developed a grazing district, and, in 1941, local landowners formed the Sheyenne Valley Grazing Association, whose goal was to improve both the productivity of rangelands and wildlife habitat. In 1954, with the grassland rehabilitation essentially completed, the U.S. Forest Service took over the administration of these lands. In 1960, 135,000 acres of native grassland, located in the sandy western portion of the delta, were designated as the Sheyenne National Grassland. About eighty families hold allotments that permit grazing on federal lands. They are assisted in their management by staff of the U.S. Forest Service and the Lake Agassiz Resource Conservation and Development Council, set up by the NRCS at the request of local farmers.

Today's grasslands are a sharp contrast to the conditions of the 1930s. Although patches of overgrazed rangeland remain a problem, rehabilitation and management efforts have transformed these blowing dune lands of the 1930s into productive pastures again.

Despite these dramatic changes, the Sheyenne Delta remains one of the most ecologically varied landscapes in all of North Dakota. A 1973 study documented more than eight hundred species of plants and wildlife in this region. This diversity is largely due to one distinguishing feature: over most of the delta, the aquifer is within ten feet of the surface when there is average precipitation. In a region that can receive less than twenty inches of precipitation a year and regularly endures some of the most extreme temperatures anywhere on the continent, this high water table supports an astonishing array of habitats. Broad stretches of prairie are interrupted here and there by wetland thickets and sedge meadows. Oak savanna grows on old dunes and choppy, sandy ground. Calcareous fens fed by springs dot the river terrace slopes. The high mineral content of this cold groundwater supports plants rarely seen elsewhere. In several delta habitats, researchers have documented plants that exist nowhere else in the state. These include the western prairie fringed orchid, handsome sedge, and slender cottongrass. The orchid populations– a federal threatened species–are among the largest in the world.

Adding to this diversity is the Sheyenne River, its tributary streams and springs, riparian forests, and woodlands that creep up the ravines and slopes. The Sheyenne Delta is a place where the tallgrass and mixed-grass prairies mingle. Its limited rainfall signifies a continental shift from a humid to a more arid region. Its woodlands are a blending of eastern hardwoods and boreal forest.

Although much of the landscape's native character remains, several land-use practices are putting the ecological richness of the delta at great risk. A combination of weed control, fire suppression, heavy livestock grazing, and cultivation pose a problem to the region's native species.

Some of the world's biggest populations of the U.S. federally threatened western prairie fringed orchid (*Plantanthera praeclara*) are found in the Sheyenne Delta. (Photo by John Challey)

The Scourge of Spurge

Leafy spurge, or in scientific parlance, *Euphorbia esula*, is a plant of small stature but enormous influence. In its baleful presence, grasslands become unproductive, most livestock find less to eat, and costs of ranching skyrocket. Arriving in the Great Plains with shipments of grain from Russia, by the early 1900s leafy spurge was well established over thousands of prairie acres. Cattle do not eat it, and pastures infested with this noxious plant become worthless. More than this, spurge forms dense patches that force native prairie plants out, reducing the diversity of a prairie. A glowing yellow cast to a prairie in late May and early June may mean that spurge has a grip on it.

Researchers throughout the Great Plains and the Red River region are striving to find ways to suppress leafy spurge in pastures and prairies. Goats and sheep will eat it; however, without careful management, their grazing can harm the prairie, and when they are removed, spurge simply springs back in full force. The most common treatment for spurge is to apply a powerful chemical, such as picloram or 2,4-D. With repeated dousing, the spurge is dramatically reduced—even killed—but so are the native wildflowers. And chemical treatment is expensive. More promising is biological control. Scientists scoured leafy spurge's native haunts in Asia to locate and bring back insects that eat spurge. Several species of flea beetles that attack roots are being released at experimental sites to ensure their effectiveness and that they will not devour native prairie species. Early results at locations outside the delta are good: the insects reduce spurge to a minor member of the biological community. Since the insects grow and spread very slowly, people must help them out by recapturing them, breeding them, and moving them to new places where their spurge-eating ways can help improve the prairie.

The yellow-green tint on the landscape is leafy spurge run amok. (Photo by Brian Winter)

Before European settlement, native prairie was renewed by two periodic disturbances: the grazing of bison and fires. Fires were set by lightning or on purpose by humans to herd animals, increase seed production, and clear the land. Fire suppression and excessive grazing, which weakens the native plants and sod and favors the spread of weeds, have led to a decline in the diversity of native species.

Weeds have replaced native plants and grasses in many stretches of the Sheyenne Delta. Efforts to control such non-native aggressors as leafy spurge, smooth brome, Kentucky bluegrass, and Canada thistle are often ineffective. The spraying of broad-spectrum herbicides to control spurge and other noxious weeds further reduces grassland diversity and adds to the cost of operations for prairie managers and owners of adjacent lands. The U.S. Forest Service estimates that at least 11,000 acres of the Sheyenne National Grasslands are degraded by leafy spurge.

At the same time, expanding potato-farming operations in the delta are making inroads into the remaining native grasslands and dividing them into smaller parcels. Problems resulting from the farming of sandy soil seen elsewhere in the country are now part of the delta landscape. For the first time since the 1930s, the wind erosion of surface soils is again a major conservation concern.

And so, too, is the availability of water. Center-pivot irrigation used primarily for potato crops, for example, can affect the hydrology of nearby springs, seeps, and fens. While withdrawing water for irrigation may not deplete the Sheyenne Delta aquifer, changes in groundwater can alter the unique habitats that make up some of the unparalleled diversity of the area.

Although private lands are increasingly being turned into potato farms, most delta residents still earn their living through beef production and cattle grazing. The Sheyenne National Grassland supplies seasonal forage for approximately 11,000 mature cattle and their calves. These herds are owned by about eighty ranch families, who are members of the Sheyenne Valley Grazing Association.

Some of the most forward-looking and far-reaching efforts to preserve the delta's unique heritage are being launched by those closest to it—the people who work the land and study its intricate functions. Here are some of their stories.

Larry Woodbury, a Sheyenne Delta rancher, lives with his wife, Judy, and their three children in McLeod, North Dakota.

My neighbor and I were out on horseback in our cattle, and he was talking about his dad, who is about ninety-four years old. His dad says that when he first came here and mowed hay with horses, the grass was taller than a horse's back. That was your native species. That was before Kentucky bluegrass invaded us.

Larry Woodbury, cattle producer, McLeod, North Dakota

Larry Woodbury and his son on their ranch in the Sheyenne Delta. (Photo by Michael Burian)

A bunch of us have been trying to manage our land more holistical-ly. I tried some things last year and feel that I've learned more in the last year about native grasses than in my whole lifetime. I'm very excited about it.

I've seen that you can graze the land fairly heavily, but then can get off it—like when the buffalo came through and grazed it—and give the land ample rest. Just putting that electric wire down the middle of my pasture and timing the grazing made a night-and-day difference from one side of the fence to the other. That's what we feel we have to do to support our native species.

With the deeper roots of these native plants, we are improving our soils for the future. We can increase the productivity of the land without as many inputs, such as fertilizer. Over the last forty years we got into a mind-set in farming that dictated, "Well, you have to use this fertilizer and you've got to buy that chemical because we've always done it this way." I feel there is great opportunity ahead. We've learned that we have to control costs and manage our grass to get the most out of it. If we can do that, we'll be way better off down the road.

A vigorous growth of warm-season grasses is evident on the left side of the fence, while, on the right, growth of the native grasses is suppressed. (Photo by Marilyn Latta, courtesy of Manitoba Naturalists Society)

Last summer I was out on horseback, and my cattle and my neighbor's cattle broke together, and we talked. He said, "Your pasture looks really good." His pasture was right across the fence. So I got off the horse, and I showed him, "You've got the same plants," but the vigor wasn't there because he hadn't given the pasture a rest. Well, his brother owned a half section that he gives a lot of rest. We rode over there, and I showed him the difference. There were a lot of native species. Just by my talking with my neighbor that one day, he saw the difference.

It's all what a guy learns. I used to go out to check the cows. Now I'm getting like some of the guys we know out West. We go out to check the pasture, the range.

It's enjoyable work. I find it very pleasing to go out there and look at the range. My wife and the kids come along with me, and Judy says she feels more a part of the operation than ever before. There's always been that love of the land, but you feel closer to it and to nature when you work with it and see these results. We realized that we probably have fallen into some ruts, that we may have hurt some range conditions, and by doing things the same way for so many years, the Kentucky bluegrass invaded and took over. But by changing the way we do things, we can get back to the way the land once was.

Tom Spiekermeier lives with his wife, Connie, on a farm near his family's homestead on the Sheyenne Delta's northwestern edge. A nature photographer, Spiekermeier is a retired farmer whose interest in nature led him to serve on the board of trustees for The Nature Conservancy's North Dakota chapter.

T. S. Eliot said, "The end of all our exploring will be to arrive where we started and know the place for the first time." I was born on a farm just east of Sheldon, North Dakota. Connie and I built a farm just off the homestead a mile south of Sheldon. We've been here all our lives. Now our son farms. We have another son and daughter with a greenhouse operation nearby.

I got interested in photography after the family grew up. Connie and I did some hiking on vacation in Itasca State Park, and we found several orchids we couldn't identify. I felt inadequate. Here we'd lived in the region all of our lives, and we didn't poke our heads out of the furrow. I went out and bought *Northland Wild Flowers* by John and Evelyn Moyle, along with a macro lens and starting taking pictures around my home in Sheldon. I was overwhelmed by the diversity I found.

This was one of the early areas of settlement in the state, with woods, water, fish, and pasture. Then in the 1930s, people farming on the sands ran into problems with the drought. The federal government came in and bought people out to relocate them. It was a sad time but also necessary—the locals say the land's been healed because it's gone back into grass.

Since then, land use shifted quite a bit. We've had grazing for some time, but now a lot of farmers are leasing land out for potatoes and carrots. This could create problems with groundwater down the road and with wind erosion. The large companies that lease the land like it because the sandy soil doesn't stick to the vegetables. Down the road, this farming might be threatening if it isn't managed right.

When you start appreciating this planet and the cleverness of nature, you reach into a spirituality of existence. This appreciation for the planet and its uniqueness creates a subtlety in your relationship to other humans. It's a humbleness, that we're only on this planet for a brief time and we need it to survive. You can talk about miracles and all the rest, but this is it—what comes after it might be considered a bonus, but this is it.

Retired farmer and photographer Tom Spiekermeier.
(Courtesy Tom Spiekermeier)

Steve Schumacher with fire fighting equipment. (Photo by Michael Burian)

Steve Schumacher is a range technician for the U.S. Forest Service. One of his duties is to help cattle grazers in the Sheyenne Delta improve the quality of their rangelands, mostly through such techniques as prescribed burning.

Burning is part of the mentality around here. We don't need to spend time convincing people that fire is okay. We put an announcement in the newspaper, and we'll get four or five phone calls from cattle operators asking us to come and burn their allotments.

We're trying to put fire back in its place in the landscape. It's one of the things that made the prairie what it is. We look at ecosystem health and try to make grass dominant in a landscape that has seen an increase in woody growth, mostly willow, buckbrush, and Russian olive. We also burn to create better habitat for the western prairie fringed orchid. By removing the litter layer, you enhance the plant's ability to survive.

We also have a goal to improve forage for livestock. Some areas of the grassland grow phenomenal amounts of forage, faster than the cattle can eat it. That's why fire is so important. After a number of years, the litter on the ground gets so thick that you end up with fewer growing plants. The plants of sedge meadows have a narrow "window of palatability" when the cattle will actually eat them. If you burn the land, you widen that window, and the cattle will eat the plants over a longer period of time. If we don't burn, the cattle use only ten percent of the available forage. After a burn, they'll use up to thirty percent.

Sustaining Diversity: Fire and Grazing Fuel the Prairie

The life of the prairie is an ebb and flow cycle of grazing and fire. Grazing keeps the warm-season grasses in check and allows sunlight to reach the seeds of short-lived plants, prompting them to reproduce and survive. It favors shorter species such as the grama grasses and little bluestem, and species that are dormant or dwarfed in the tall, dense grass of mature prairie. In areas of short grass, greater prairie chickens dance and mate, and upland sandpipers whistle and nest.

Prairie and fire go together like a car engine and gasoline. A prairie runs better with a fire every four or five years to stimulate growth of the grasses, make the wildflowers bloom, and knock back weeds and brush. Burning and stripping away the litter exposes the prairie surface to sunlight, stimulating new growth and revitalizing plants and the soil. Phosphorus returned to the soil in ash causes a profuse blooming of plants. After a fire, the prairie comes to life: insects return, small mammals dig new burrows, hawks hunt the burn site, and some dormant seeds germinate.

In the way that peaks and valleys in the economy create financial opportunities, grazing and fire together renew conditions favorable to many prairie plants and animals that fail to thrive in tall, dense growth or deeply littered ground.

Sighting down the fireline of a controlled burn. The mowed break is wetted down by an all-terrain vehicle fitted with a hose and pump (in background). (Photo by Brian Winter)

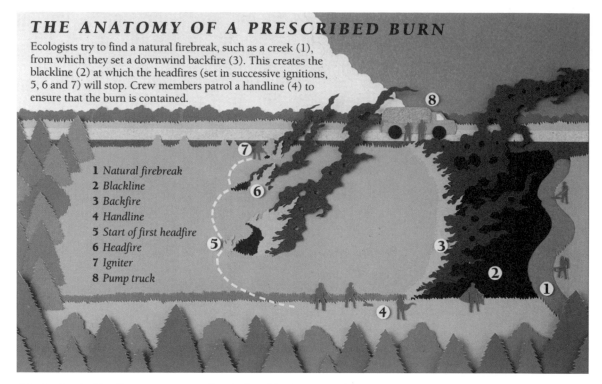

THE ANATOMY OF A PRESCRIBED BURN

Ecologists try to find a natural firebreak, such as a creek (1), from which they set a downwind backfire (3). This creates the blackline (2) at which the headfires (set in successive ignitions, 5, 6 and 7) will stop. Crew members patrol a handline (4) to ensure that the burn is contained.

1 *Natural firebreak*
2 *Blackline*
3 *Backfire*
4 *Handline*
5 *Start of first headfire*
6 *Headfire*
7 *Igniter*
8 *Pump truck*

(Illustration by The Nature Conservancy)

Dr. John Challey, a retired biologist, is a longtime researcher in the Sheyenne Delta and member of the board of trustees for The Nature Conservancy's North Dakota chapter.

Upland sandpiper (formerly called upland plover). (Photo by John Challey)

I've always been interested in the area, back to the days when I hunted pheasant with my parents. I remember being there on opening day in 1939. When the sun rose, it was like a war down there. We were staying at the 4-H camp. At that time, there were a lot of abandoned farmland and a lot of weeds, which made for good pheasant habitat.

I did my master's thesis in the mid-1950s on white-tailed deer. When I started, the area had been managed by the Soil Conservation Service for about ten years. It was purchased by the Resettlement Administration, a New Deal agency. About 1954, the U.S. Forest Service took over and made some relatively big changes. The agency instituted rotational grazing, which made quite a big difference. Back in the old days, there was very little cover left on the land. Now there's more native vegetation—and a greater diversity of species. People in Minnesota are looking for upland plovers, but here in the Sheyenne Delta, they're as common as meadowlarks. When I hear that the regal fritillary is a rare butterfly in Minnesota, it surprises me because it's very common here. This year we had a wet spring, and the orchids were showing up all over the place. They might bloom some years, then not bloom again for ten or twenty years. All of a sudden, you get the right conditions, and there are thousands of them. When Alexis Duxbury, a Natural Heritage botanist with the North Dakota Department of Game and Fish, first did orchid counts in 1984, she found 2,000 plants. Now there are tens of thousands of documented plants. Unfortunately, many pastures are still overutilized. There's been an effort by the Game and Fish Department to influence the Forest Service to reduce grazing around prairie chicken booming grounds.

I'm personally interested in the riparian areas, which dates back to the time when I was chasing deer. In 1982 I started volunteer survey work for North Dakota's Natural Heritage Program. I was interested in the fact that there are plants of northern bogs and the eastern woodlands found in the riparian areas. The Forest Service has done some riparian protection over the years to exclude cattle from certain areas, and hopefully that will continue.

Regal fritillary (*Speyeria idalia*), a large butterfly of tall-grass prairies. (Photo by Robert Dana)

Grazing Periods for Pastures

Cool Season Grasses 15 June–15 July	⊶⊶⊶ Electric Fence
Cool and Warm Season Grasses 15 May–15 June 15 August–1 September 1 September–30 September	〰〰 Watering Hole
	Wetland
Warm Season Grasses with Brush 15 July–15 August	Dry Ridge
	══ Road

1/2 mile

Bettering the Ecology and Economy

The Lake Agassiz Resource Conservation and Development Council (RC&D) was formed by local citizens using a program of the U.S. Department of Agriculture. Designed to improve rural economic and community development through the stewardship and betterment of natural resources, the Lake Agassiz RC&D undertakes projects in a multi-county area of southeastern North Dakota.

In 1996 several ranch couples in the Sheyenne Delta formed the Sheyenne Delta Holistic Resource Management Workgroup. They meet regularly to learn and share their experiences in applying a holistic approach to their operations. By making better informed management decisions, these couples hope to improve their quality of life while boosting the diversity and productivity of their prairie pastures. RC&D staff actively support this workgroup by helping to sponsor workshops on holistic decision-making, biological monitoring, and the creation of wealth.

Ranchers can maintain or improve conditions by moving cattle into a pasture at a time when grazing does the most good.

Jay Mar is coordinator of the Lake Agassiz Resource Conservation and Development Council based in Fargo, North Dakota. Mar grew up on a ranch in central North Dakota.

It's amazing how well the ranchers who practice holistic resource management know their land. They can tell you whether they have this species or that and pick out the exact hill, the low area, or the quarter where it's growing.

That kind of attention is really important when you're implementing timed grazing. The producers watch the bite of the cows to monitor how much of the plant the cow is using during grazing. They don't want the cow to bite the plant too low and damage the growing point. By doing that, they're getting more in tune with plant physiology. You might say they *are* the botanists. There are plenty of producers out there who have the potential to be the botanists and ecologists in the field. It is just beautiful when you recognize this potential and their yearning to work together. That is their life. They are out there every day in those pastures.

What's come out of the many holistic resource management meetings I've attended is an identification of values that are important to people in rural areas: these values are family, the land, and their environment. I believe that every species in nature has its place, and I do not believe that we should lose any of them. I believe that the family farm is something we also need to preserve. We need both for the future.

In the many meetings we have had around issues of holistic resource management, the message is the same: The values important to people in rural areas are family, the land, and their environment. That's basically what people use to do goal-setting around here: the family and concern for their environment, taking care of the land.

Jay Mar, coordinator,
Lake Agassiz Resource Conservation and Development Council, Fargo, North Dakota

Ranchers talk about soils during a Summer Range Tour in the Sheyenne Delta sponsored by the Lake Agassiz Resource and Conservation Development Council. (Photo by Jay Mar)

Chapter 5

Living with the Prairie
In the Agassiz Beach Ridges

I took my wife out to a friend's blind last spring, first time she'd ever seen a prairie chicken. The males come in stomping up a storm, trying to attract the females that just walk right by them, hardly giving them the time of day. Jan started laughing so hard, I thought I was going to have to gag her.

It's important to get people out on the prairie. This is what grabs them. Visiting a lab doesn't do it. Walking them through the prairie and telling them a few stories really gets them excited.

Gerald Van Amburg, prairie ecologist, Concordia College, Moorhead

Ralph and Roberta Ouse are retired farmers living in Rothsay. Their son continues to hay the native prairie on their land while nearby the prairie chickens boom each spring. Art Fosse (seated) designed and built the roadside prairie-chicken, with support from the Ouse's and others on the community's bicentennial committee. "The prairie chicken has put us on the map." says Roberta. "The truckers sometimes call us the Big Turkey, but they know where they're going. It's a landmark. We have the sandhill crane here too, in large numbers in the fall." Ralph continues, "You can tell when they're going to leave; you'll hear on the forecast that a cold front is coming, and the last nice sunny day they'll be circling in the sky about as high as you can see them, and the next day they'll be gone." (Photo by Michael Burian)

Gazing westward from the low hillcrests of northwestern Minnesota, early settlers often described the tallgrass prairie as a great inland sea, its grasses undulating in the wind like ocean swells as far as the eye could see. Little did many of them know that they were poised on the eastern edge of an ancient lake, about to embark across the plain of a vast lake bottom, the dips and rises in the land a series of gentle sandy terraces shaped by the waves of Glacial Lake Agassiz.

These scattered low-rising ripples in the land, known as the Agassiz Beach Ridges, extend in a north-to-south band across five counties. In some places, they have been erased by repeated plowing and leveling of the landscape. One way to see the ridges today is to drive east on U.S. 10 out of Moorhead, through Dilworth and past Glyndon. Near the entrance of Buffalo State Park, the land rises slightly. This gentle slope is the ancient Campbell Beach, the second youngest beach to be formed by Glacial Lake Agassiz. It was identified in 1881 by geologists Warren Upham and Horace Winchell in their quest to locate the principal beaches that marked the lake's

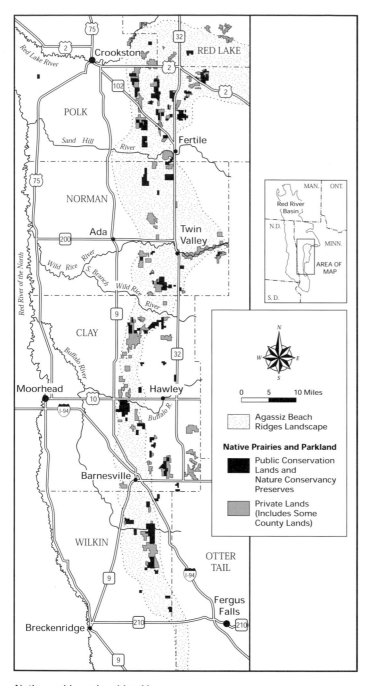

Native prairie and parkland in the Agassiz Beach Ridges Landscape. (Source: The Nature Conservancy Great Plains Program and Minnesota Natural Heritage and Nongame Research Program)

shores. A few miles beyond, on the hill to Hawley is the higher Herman Beach, described in an 1888 government report as "a broad wave-like swell, with a smooth, gracefully rounded surface. Here and there were gaps cut through by streams." The ridges twist for miles north and south along the east margin of the Red River Valley, with several lined up one behind the other in some places, such as southeast of Crookston.

About 10,000 years ago, as the waters of Glacial Lake Agassiz drained southward, the lake's eastern shore was little more than a series of sand terraces and rock-strewn ridges rising from the flatland valley floor. By about 8,000 years ago, tall grasses and flowering plants covered the land. Animals, insects, butterflies, and birds flourished, each species evolving its own adaptations to the nuances of moisture, temperature, and light found in the varied mosaic of habitats. In time, humans came to hunt wildlife and gather plants for food, medicine, and ceremonies.

Life on the beach ridges was as abundant as the land was spacious. The region once supported huge, continuous populations of animals. Bison and elk and their predator, the wolf, ranged freely between the beaches and adjoining prairie landscapes. Flocks of waterfowl, sandhill cranes, and other birds moved through in very large numbers, as they do today. The variety and abundance of plants and animals were shaped by the prairie fires, which sped through the landscape, leaving growth and renewal in their wake. Observing that lightning fires drove herds of animals in predictable directions, early American Indians began to use human-sparked fires as a hunting tool.

From the vantage of the beach ridges, these hunters plotted their strategies. Later, in the nineteenth century, drivers of the Red River oxcarts used the beach crests to survey the best routes for transporting trade goods between Winnipeg, Pembina, and St. Paul. Traders met to transact business and socialize at places such as the Hudson's Bay

Bison hunt, Metis Indians and white settlers, near Pembina. (From *Harper's Weekly*, October 1860, Courtesy of Minnesota Historical Society archives)

Making Hay While the Sun Shines

A haircut is a rejuvenating experience for most of us. We look a little spiffier, walk with a jauntier step, and those wayward locks don't fall in our eyes. A prairie can also benefit from having its old top removed from time to time. In the upper Midwest, prairie hay should be cut after young birds leave the nest—sometime after July 15—but before the grasses put on their growth spurt to produce seeds in late August and September. The hay provides a lower-cost forage for range-fed cattle at stockyards, for dry cows, and for horses; strawberry growers and landscapers also use it as a weed-free mulch.

A farmer who is treating his or her prairie well will take the hay bales off the prairie right away, not use heavy equipment that can break the sod, and clean the equipment before haying so that the seeds from weedy exotic plants—such as sweet clover, thistle and leafy spurge—don't invade the prairie. Most farmers hay their prairies every year, but a prairie will benefit from a rest so that the native plants can produce and spread their seeds to produce young plants again.

A cared-for hay meadow can appear like a beautiful flower garden in June and July because haying, like a mid-summer fire, suppresses the vigorous tall prairie grasses that can crowd out plants that flower earlier in the year. Hay meadows also provide a dance-floor for greater prairie chickens and sharp-tailed grouse to strut their stuff, and some types of prairie birds—marbled godwit, for instance—prefer to nest in the short grass, unhampered by the stalks of past years' growth.

Haying helps control non-native plants like sweet clover, on prairie grasslands. (Photo by Brian Winter)

False asphodel (Tofieldia gluti-
nosa), a plant typically found
in fens and calcareous wet
prairies within the Red River
Valley. (The Nature
Conservancy photo archives)

Company post at Georgetown, Minnesota, operated by R.M. Probsfield, who is considered the first settler to occupy land in Clay County.

Probsfield may have been the first person to break the prairie soil of
this region, or perhaps it was his friend David McCauley, who seeded
seventy-five acres of oats north of Breckenridge in 1861. Today, the
vast native grassland that settlers viewed with admiration and some
trepidation has been almost entirely converted to farmland. But here on
the sandy windswept swells and in the marshy troughs of the beach
ridges are some of the largest and best northern tallgrass prairie habitats, including calcareous fens, wet meadows, moist and dry grasslands, and oak-aspen savannas. Rare orchids, butterflies, and birds
thrive in these habitats as they did thousands of years ago. In prairies
like these across Minnesota, one-third of the state's threatened or
endangered plants and animals still find a home.

These prairie remnants exist on marginal lands whose rocky
soils damaged farm machinery and whose wet soils discouraged the
plow. Since European settlement, the beach ridges have been used
for haying and grazing, but earning a living has often been difficult,
and widely fluctuating land values have added to the economic
uncertainty. Today's livestock producers maintain cattle, sheep, and
a few bison. Increasingly, these landowners are turning to gravel
mining as an additional source of income.

Unfortunately, bountiful gravel quarries operate in what are the last
best habitats for prairie chickens, cranes, small prairie mammals, rare
plants, and butterflies. Day after day, trucks haul hundreds of tons of
construction material from these ancient lake terraces to construction

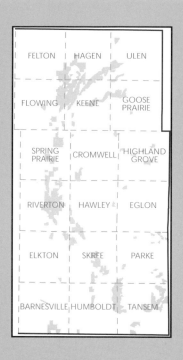
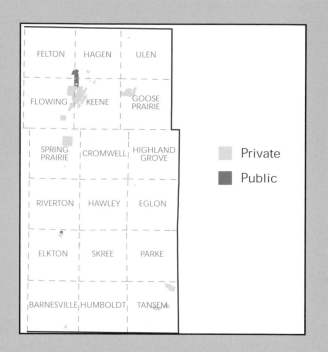

Private

Public

A Prairie Balancing Act

From 1995 to 1997, representatives of the gravel industry and of county, state, and federal agencies, together with local officials and private landowners, met and discussed the future of prairie and the gravel industry. These twenty-seven people, coordinated by Cindy Buttleman of Minnesota's DNR's Mineral Division, realized that the best hope for preserving prairie and providing much-needed gravel for a growing economy was to talk about it in a neutral setting before these resources come into conflict. Using Clay County as their experiment, the group gathered information on prairie and gravel, turned it into computerized maps and other easy-to-use data, and made them available for community leaders and citizens. The group also recommended the best ways to reclaim old gravel pits scattered throughout the Agassiz Beach Ridges.

Location of high and medium-quality prairie in Clay County (left) and prairies with high potential to contain gravel and sand deposits (right). (Map by Minnesota Department of Natural Resources, Minerals Division)

sites in the cities expanding along the Red River. The mining destroys prairies and adds to the pressures on existing preserves. And some observers worry that mining activities pollute and deplete the underground reserves of water that feed many rare prairie habitats and supply drinking water for thousands of rural residents.

The Agassiz Beach Ridges are the focus of several collaborative projects designed to measure the effects of mining on the prairie ecosystem. For example, CAMAS, an international mining corporation, supports the ongoing assessment of the impact of mining on the hydrology of the beach ridges, in particular its impact on one of the largest fens in Minnesota. CAMAS has a special interest in the monitoring project since it removes aggregate from an eighty-acre site leased from the State of Minnesota within the Felton Prairie Complex where the fen is located. The assessment, conducted by the Minnesota DNR, began in 1995 with the drilling of a series of deep wells for measuring the water levels and the chemistry and quality of groundwater, and a shallow well for tracking surface-water fluctuations.

Roy Sander, production manager, says that the CAMAS corporation's mission of stewardship, community service, and sound business practice is reflected in this project. "We want to find out if mining activities are artificially lowering the water level," he observes.

Another gravel mining/prairie-protection study, funded by the Legislative Commission on Minnesota Resources, is developing a conflict-resolution approach to allow for continued gravel mining while protecting prairies. In a related effort, The Nature Conservancy is working on a six-year prairie-reconstruction plan for several abandoned gravel pits in the Agassiz Beach Ridges to develop the best methods for repairing grasslands. The plan is based on the premise that planting prairie species is a practical way to curb erosion on open land and mimic natural processes. However, prairie ecologists warn that it will take at least a hundred years—if ever—for a fully functioning prairie to reestablish itself. Finding ways to locate new mine sites away from rare habitats, they say, is a surer way to conserve native prairie than rehabilitation.

The Nature Conservancy, government agencies, and area gravel operators are just a few of the groups investigating new ways to strike a more habitable balance with the precious prairie resources of the Agassiz Beach Ridges. They have joined area residents from all walks of life to ensure that these remnants of our heritage, which reach beyond the region's pioneer past to its glacial beginnings, survive for generations to come.

Brad Bjerken, owner of the B Bar B ranch in Felton, Minnesota, lives with his wife and their son on his family's land. They have a cattle operation as well as a gravel mine on their property.

Brad Bjerken, rancher and conservationist, Felton, Minnesota. (Photo by Michael Burian)

Gravel pit at Bluestem Prairie, Glyndon, Minnesota (top), 1985. (Photo by Margaret Kohring)
Bottom: Restoration in progress, same pit in 1991. (Photo by Brian Winter)

My grandpa bought this land in 1956. We later incorporated it as a family corporation. We operated a wool pool here—raised, graded, and shipped lambs. We lambed up to 4,000 sheep. During certain parts of the year, we had to have twenty-four-hour lambers.

The last sheep left the place in 1973 because the land seemed better for cattle. We have about 250 beef cattle and just about as many calves. Beef prices are really low now, about half of what they were a year ago. I think it's due mostly to overproduction. When people saw the prices go high, everybody raised more cows. Even though beef prices are low, the feed prices are high, especially oats and barley. There's not as much corn grown for cattle in the valley as there used to be.

We have 3,743 acres altogether; of that, 2,166 are in native prairie. First thing in the spring, pasque flowers are all over the high sandy ridges. Then other stuff comes. In August and September, I can just be amazed at all the different plants I see growing. I can see so much.

A lot of people look at a piece of land and say, "It's just grass," and I can tell them, "It's weed-infested, overgrazed, and looks terrible." I can tell native prairie when I see it. It's different from the other.

In the past, some of this land was overgrazed. Now we rotate the cattle from pasture to pasture. I can use the pastures that have cool-season grasses in the spring, then get the cattle off and let the grasses come back. That way, I can graze my cattle again until snow covers the ground, usually into October and Novem-

Prairie hunters (Blacksmith Al Travnicek, at right, with son Leo) and over fifty ducks after a day's work, circa 1920, Hawley, Minnesota. (Photo by S.P. Wange, Hawley, Minnesota, Courtesy of the Flaten/Wange Collection, Clay County Historical Society)

79

ber, compared to most people around here, who start feeding their cattle in September because their mid-season grasses are all used up by then. We get several more weeks of grazing than if we just let the cows pasture continuously all over the place.

I can tell you what I want this land to look like in thirty years. The gravel pit would not be one foot closer to the house. And all the native prairie would still be here, just like it is.

We've got a good deal going. We're trying to mine only the deep gravel. That way, less of the prairie is disturbed. The big equipment that rolls in and out of the site can do awful damage. It packs the soil, really wrecks it, and then weeds take over. It's terrible. The deeper mines do less damage than shallow deposits. This pit is going to last a good long time.

Russell Fitzgerald is a retired railroad worker who lives in Dilworth, Minnesota.

When my dad was a kid, they'd take a load of grain from the Sabin place to the mill at Stockwood, east of Glyndon, about ten miles. While the wheat was being ground into flour, they'd go fishing in the Buffalo River. They really enjoyed that. That was back in the 1800s. I remember when there were more streams coming off that area called the Beach Ridges. Many people had what they called "running wells," or springs, coming right out of the ground.

Paul Kaste of Fertile, Minnesota, sits amid sacks of native prairie seed he grows for farm lands enrolled in the Conservation Reserve Program. (Photo by Michael Burian)

Hunters came from Minneapolis to Sabin. They'd get off the train, rent a rig, come out to my grandpa's place, then transfer to a springboard wagon and load it up with prairie chickens, then get on the train and go back to Minneapolis.

Special train cars brought in the big shots who liked to hunt. They'd pull their coaches and dining car off onto Richard's spur, a sidetrack here in Dilworth. From there, they could walk a distance of about six blocks to shoot ducks and prairie chickens.

We had a season we called spring hunting, when the land between Sabin and Moorhead was flooded. It would sometimes be full of ducks and geese.

During World War I, a lot of land around here was plowed up for farmland; yet, there was still enough open space for the prairie chickens to prosper. They also liked the corn-fields, especially the shocks of corn we stood up in the fields in the fall. The chickens used them for cover.

When I was a kid, there was a sixteen-a-day limit on prairie chickens. They were good to eat. You could almost taste the prairie from all the seeds they ate. There is a small covey of them here in Dilworth. At nightfall I see them go into the prairie grass along the railroad tracks and back again in the morning.

Paul Kaste, a native-prairie seed grower, lives on the farm on which he was raised near Fertile, Minnesota.

I attended the University of Minnesota in St. Paul and graduated with an agronomy degree in 1959. I worked outside of farming for about twelve years for the Production Credit Association. I always maintained my homestead here—bought it from my parents—and I always worked with the idea of having a life as a farmer. In 1974, I was able to go into farming full time.

I am the third generation to farm this marginal land. We're on the east side of the county where it's not really productive for farming. Where valley farmers might get fifty bushels to the acre, I'll produce twenty-five or thirty.

I've been in the seed business for almost ten years. I started with the help of the Soil Conservation Service in Bismarck, North Dakota. They furnished me with cultivars, and they gave me advice. I'm still using that material. Cultivars are native seeds that have been improved. SCS is primarily interested in trying to produce something for agriculture rather than for restoring prairies. But in gathering material from many different locations and by observing, they would find plants that show more seeds or leaves for good pasture, and we decided to produce them. I started with bluestem and was apprehensive at first. But it worked fine, so I added switchgrass and then Indian grass.

The big demand came when the Conservation Reserve Program was enacted eleven years ago. There never was any way to supply enough native seed for the thirty-two million acres in that program. Most landowners seeded their CRP land with such grasses as timothy and brome. But it did stimulate the grass seed market, and prices shot up. There was just no way to meet that demand. A single customer could buy everything I could produce.

(Editor's note: The 1996 Farm Bill increased demand for native seed plantings on CRP, and seed prices have risen again as a result.)

Dr. Gerald Van Amburg is a prairie ecologist at Concordia College in Moorhead, Minnesota. He has done research on grassland ecology. More recently, he has studied water management, particularly the role of wetlands in watersheds and watershed management.

Gerald Van Amburg, professor and prairie ecologist, at Concordia College, Moorhead, Minnesota. (Photo by Michael Burian)

What the Agassiz Beach Ridges, Sheyenne Delta, and Tallgrass Aspen Parkland have in common are vestiges of native vegetation and the associated animal species much as they were before the areas were settled by Europeans. We just don't have very much of that left.

They also offer us places to study how natural ecosystems operate. These areas are gene banks—literally—of material that may be useful somewhere down the road. We know a lot about them. But we are finding, as in any scientific endeavor, that you keep finding more and more that you don't know. We keep making new connections and coming up with new ideas. We would be foolish if we didn't recognize that.

One of the things we understand least about native prairies is the soil and what goes on in the root area. It's hard to study. You're talking about very small things: bacteria, nematodes, and microbes. We're just starting to appreciate their mutually beneficial relationships with plants and plant roots.

One thing I fight all the time is getting people to recognize the public values of native prairie. For example, native prairies do not contribute to the maintenance costs of drainage ditches.

Nematodes and other organisms help produce healthy prairie soils. (Courtesy of University of Minnesota Press/Wadsworth Press)

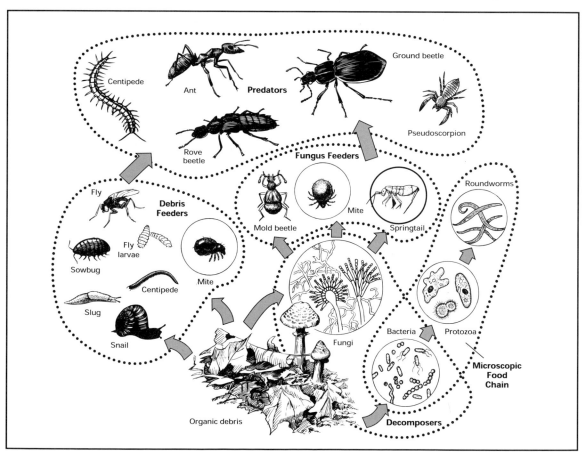

They tend to hold a lot more water back than cropland. The prairie preserves lie within the same landscape as all the farmland around it. Farmers have to pay for the drainage ditch that funnels water off agricultural land. They have to pay to keep the ditch in good shape. Residents say these prairie preserves should be taxed, and I don't deny that there should be a tax on them. But I would argue that they should be taxed at a much lower level because they aren't contributing water to the ditch system. If anything, they're holding the water back.

This region is truly a food basket of the world, with just priceless soils. And where did those soils come from? They came from the prairie. It would be a terrible shame to lose the ecosystem that provided us with those soils.

We have such a soil resource here that we almost take it for granted and don't recognize that soil depletion is a huge, huge problem. Just look around the world. We've lost some very productive spots. Agriculture started in the Tigris-Euphrates valley in Mesopotamia–gone now. It was wrecked, probably most of the problem due to irrigation and erosion. The old biblical trees they talked about aren't even there anymore. What you have now is a region of limestone scarps that will never have soil on them again, at least not in this geological time.

Family picnic in the great outdoors. High-quality food depends on good-quality soil, such as that formed by prairies. (Photo by Don Breneman. Courtesy the University of Minnesota Extension Service)

"At the same time, we have more and more people to feed and less and less productive land to do it with. We're losing it to urban development. We're losing it to roads. And we're losing a lot of it to soil erosion and salinization.

"And maybe it's because I'm getting older and more philosophical, but I'm starting to ask myself the question, What right do we have to destroy these soils and other forms of life? What gives us the authority to do that? Millions of years of evolution, and we just wipe them out. That's a cavalier attitude that I don't think we should have at all."

Sheep ranching provides an alternative to intensive farming on fragile grasslands. (Photo by Don Breneman. Courtesy the University of Minnesota Extension Service)

Tim Hagle is a rancher who lives in Red Lake Falls, Minnesota. He grows a diversified crop of wheat and corn in addition to grazing sheep.

At forty years old, I'm considered a fairly young farmer. My uncle passed away, and I took over the land he farmed with my father. I have approximately 400 acres, and 270 acres are in grass and hay. I don't believe any of the grass is true prairie on our farm because, when I purchased it, it had been tilled and plowed.

I keep a lot of my land in grass because it's fragile, and I don't much care for chemical-type farming. I've done well enough by not using chemicals. When my grandfather owned this land, he didn't use chemicals. He rotated crops and did whatever would make them the most money to live on. My grandfather believed in tillage, and he always had livestock and raised hay and natur-

84

al grasses. My father and uncle did too when they had small acreages. The machinery was smaller, and you weren't able to work up as much land. Nowadays, with larger acres and larger machinery, you can work a quarter of the land in a day. In those days, we worked as soon as the snow was gone in the spring and didn't stop until the snow was a couple of inches deep in the fall.

My father started with no chemicals. Then in the 1960s and early 1970s, they started using small amounts of chemicals. Chemicals helped them get things done sometimes, but it is a short-term solution. It's like putting a Band-Aid on your finger if you cut it off. The weeds are still there, and they are becoming more resistant. You have to give them bigger and bigger doses. In the last twenty-five to thirty years, I've seen the amounts escalate, if not in larger amounts then in more and more powerful chemicals just to get the same results. It used to cost pennies, but now it costs sometimes between twenty and thirty dollars an acre. Using all those chemicals doesn't really increase the farmer's bottom line.

Many people don't know how to farm or do things on their land without the chemicals. They can't believe that anybody ever could have farmed without them. People used to do it all the time. Of course, some guys did well and some didn't. But that's always true. Some people are good farmers and others aren't.

I have a better opportunity to make a long-term living with sheep and grasses. If I were going to have a conventional monocrop, chemicals might be the best way to get me another twenty or thirty years out of the deal, but if you're looking more years ahead, I'm not sure how long that can last.

With grasses and forbs that come back every year, I've eliminated the expense of thirty to forty dollars an acre that I would have spent planting crops. Right now I can net about $100 an acre by growing livestock on those acres. Today, unless I'm growing sugar beets or maybe potatoes on a year-in, year-out basis, I can have a more reliable income than in conventional agriculture trying to raise crops with high tillage, fuel expenses, crop insurance, and all the rest. With grasses, I don't need the same insurance: it can hail a ton and it's not going to kill the grasses.

Chapter 6

Living with the Prairie
In the Tallgrass Aspen Parkland

We have received great gifts as our heritage in Canada—beautiful rivers and parks, rolling hills, trees and grassland; the rich resources of nature are ours to protect, cherish and enjoy. Wherever man has been, there are marks of defilement and destruction—let us resolve that this shall not be our legacy to the twenty-first century. Let us rather honour and emulate those who went before us as pioneers, who gave of themselves, building and creating, enriching the lands. They have earned our thanks, and while we cannot repay them, we can respect their achievements and resolve that we will try to follow them in action and in attitude. For this we set aside this special place in our Province, a site that is part of history.

From a public information booklet published by St. Michael's Ukrainian Greek Orthodox Church, Gardenton, Manitoba

According to ecologists, the greatest variety and abundance of life can often be found at the crossroads, the in-between places where different habitats come together to form the busy boulevards, marketplaces, noisy nurseries, and quiet retreats for hundreds of living creatures. Look for the edges, they say, the places where water meets land, where rich soils are bounded by gravel ridges, where grasslands border the trees, where the opportunities to hunt and hide, feed and breed, nest and rest are multiplied by the sheer diversity of plants and animals—places like the Tallgrass Aspen Parkland.

The wind-tossed expanses of the Tallgrass Aspen Parkland remain one of the wildest outposts of the tallgrass prairie ecosystem in North America. (Photo by Robert Dana)

Native prairies and parkland in the Red River region. (Source: The Nature Conservancy Great Plains Program, Manitoba Conservation Data Centre, and Minnesota Natural Heritage and Nongame Research Program)

A unique, biologically rich constellation of natural systems forms the tallgrass aspen parkland, a place where the prairie meets the northern peatlands and aspen woodlands. Stretching in a northwesterly arc from northern Minnesota to the foothills of the Rocky Mountains in Alberta, Canada, the landscape takes its name from the open, parklike appearance of grasslands dotted by shrub thickets, wetlands, and groves of aspen.

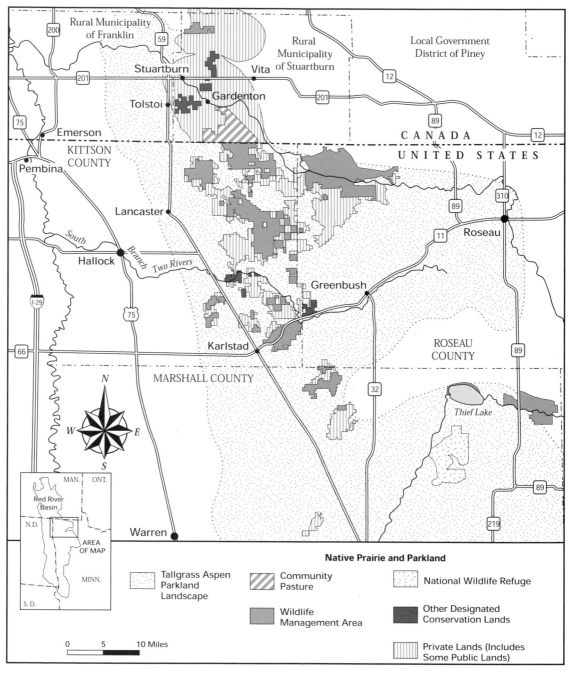

The most easterly portion of the parklands, from the northwestern tip of Minnesota through southern Manitoba to slightly north of Winnipeg, is known as the Tallgrass Aspen Parkland. This ecosystem is unlike any other in the world.

Unlike the drier parkland regions to the west, the Tallgrass Aspen Parkland is set apart by its abundant Indian grass and big bluestem. Receiving greater rainfall than the westerly plains—more than twenty inches of precipitation annually—the prairies here are characterized by stands of these rippling grasses.

But the name is deceptive. The tallgrass aspen parkland is far more complicated than the simplified system of grasses and trees found in many urban parks. From the air, for example, the tallgrass aspen parkland is an intricately fit mosaic of sedge meadows and brush prairies, wetlands and sand dunes, aspen groves and willow swamps. From the ground, views extend in every direction, making this northern region the land of the long horizon. The tabletop terrain is broken here and there by tufts of willow or aspen trees and by the undulating horizon of ancient beach ridges that rise no higher than twenty feet over the plains. Unlike the prairie-woodland ecotone in southern Minnesota, where the moraine often creates an abrupt transition, the Tallgrass Aspen Parkland occupies a broad zone with little topographic relief.

For decades, this place of subtle, intense beauty has been the province of grain growers, small livestock farmers, and hunters. Lacking spruce-rimmed lakes or dramatic topography, the Tallgrass Aspen Parkland has not received as much attention as other regions. But in the past decade, the public, as well as natural resource experts, have come to recognize its unique character and ecological value. This land is a bird-watcher's paradise, harboring more than 100 species of birds, many of them migratory, whose populations are waning as their nesting sites disappear in northern latitudes. The mix of prairie, wetlands, and brush makes the tallgrass aspen parkland one of the best habitats for spotting such birds as the short-eared owl, sharp-tailed sparrow, yellow rail, marbled godwit, and Wilson's phalarope, all listed as species of special concern in Minnesota.

The short-eared owl (above) and marbled godwit (below), species of "special concern" in Minnesota, nest in the Tallgrass Aspen Parkland. (Illustrations by Vera Ming Wong)

Populations of the Wilson's phalarope (top, left) and yellow rail (top, right) have declined in most of the tallgrass prairie. (Illustrations by Vera Ming Wong)

Cosmopolitan elk move across the international border between Minnesota and Manitoba. (Manitoba Ministry of Natural Resources)

Roaming the Tallgrass Aspen Parkland also are some of North America's largest mammals, including the elusive cougar, lynx, timber wolf, and black bear. The region harbors Minnesota's second largest population of moose. Elk from Manitoba recently have begun spending time in Kittson County, Minnesota.

And during migration periods, the Tallgrass Aspen Parkland serves as a bustling highway for shorebirds and waterfowl. Perhaps the most unforgettable visitors to this region are the sandhill cranes. As Robert Dana, a member of the Minnesota County Biological Survey team, points out, "One of the most intense pleasures of this landscape that's not available in many places in Minnesota is the presence of the sandhill crane, particularly in the late summer and fall when there's a massive staging for migration. You could just spend the day sitting out there listening and watching these birds. When you're out in this country in the spring, you're never out of hearing range of a nesting pair of cranes." Their liquid bugling and fleet, almost prehistoric profiles are reminders of the ancient rhythms that still govern life in this stark, beautiful place.

Sandhill cranes (above) and moose (below) benefit from large tracts of tallgrass aspen parkland habitat. (Photos by Henry Kartarik [cranes] and Barb Coffin [moose]. Courtesy the Minnesota Department of Natural Resources)

Our awareness of the beauty and biology of the Tallgrass Aspen Parkland is growing. In 1990, a team of scientists from the Minnesota DNR's County Biological Survey took a closer look at this subtle landscape in Minnesota and painstakingly documented its extraordinarily rich and interesting inhabitants. Their inventory revealed 815 new locations for rare plants and animals, and provided dramatic new documentation of the area's rare birds. Three years earlier, the Manitoba Naturalists Society conducted a survey of tallgrass prairie in Canada and located large prairie remnants near the towns of Tolstoi and Gardenton just north of the Minnesota border in southeastern Manitoba. In 1989, the Manitoba Critical Wildlife Habitat Program, a cooperative initiative involving seven conservation organizations, began consolidating these tracts in what is known as the Manitoba Tall-Grass Prairie Preserve. These remnants harbor more than thirty species of butterflies and 140 species of birds. Among the 300 varieties of plants is the endangered (in Canada) western prairie fringed orchid. The preserve is the only place in Canada where the orchid's feathery blossoms still grace the prairie summer.

These surveys made significant inroads in our understanding of the unique character of the Tallgrass Aspen Parkland. But identifying the species in this ecosystem is just the beginning. Despite these dramatic new findings, surprisingly little is known about the larger interactive forces of climate, topography, and natural disturbance

The Manitoba Naturalists Society logo.

Bur oak savanna at Twin Lakes Wildlife Management Area, Kittson County, Minnesota. (Photo by Robert Dana)

92

that sustain this forest-grassland ecosystem. Reconstructing a history of the region's vegetation is difficult. The region lacks lakes and undisturbed peat bogs from which scientists can extract sediment cores and read the layers of trapped pollen that record vegetational changes over hundreds of years.

Researchers have had to rely on eyewitness accounts from nineteenth-century expeditions. In 1878, surveyors described the Minnesota tallgrass aspen parkland as low prairie covered with lush grasses and willow thickets. In *The Palliser Expedition*, a chronicle of the mid-nineteenth-century exploration of the Canadian prairies by the Irish adventurer John Palliser, author Irene M. Spry describes the prairie-woodland ecotone Palliser saw from a high point along the Pembina River: "To the east the country was wooded and irregular; to the west, at the higher level, there was nothing but bare prairie lands."

Judging from early reports like these, one thing seems certain: without the frequent disturbances on which the tallgrass aspen parkland has come to depend, the ecological balance has tipped in favor of the aspen tree, which is quickly closing these open spaces and transforming them into forest. Before its conversion to agriculture, the tallgrass aspen parkland relied on sweeping wildfires, which native inhabitants called the red buffalo of the prairie. These

Wildfire on the banks of the Red River. (Illustration by William A. Rogers, from *Harper's Weekly*, December 1878, courtesy Minnesota Historical Society archives)

Aspen grove, containing remnant plants of prairie habitat, Kittson County, Minnesota. (Photo by Robert Dana, courtesy of the State of Minnesota Natural Heritage and Nongame Research Program)

fires checked the growth of trees, as well as released nutrients into the soil. The grazing of tallgrass aspen parkland vegetation by bison and elk also helped control the spread of woody vegetation while providing the added benefits of fertilizing the soil and dispersing seeds. In the absence of these agents of change, aspen trees have encroached on the grasslands of this prairie-forest ecosystem, leading some ecologists to declare it one of North America's most endangered ecosystems. Left unchecked, aspen growth could have a devastating impact on the tallgrass aspen parkland ecosystem. Scientific evidence suggests that since the 1880s, aspen woodlands have greatly expanded their range into grasslands. In some places, researchers have found that tallgrass aspen parkland evolved into aspen forest in little more than two decades.

The most vital ingredient for restoring and maintaining the tallgrass aspen parkland ecosystem is fire. Effective fire management depends on large blocks of contiguous land. Ensuring that these blocks are kept whole is vital to their ecological health. The plowed fields, road construction, vacation homes, and human settlements that checkerboard the tallgrass aspen parkland have made fire management more difficult, risky, and costly, sometimes ruling it out altogether. When fires do break out on wild lands, U.S. agencies

such as the Minnesota DNR's Forestry Division are under mandate to extinguish them. Some fire-suppression methods, such as bull-dozing smoldering peat burns, have damaged fragile habitats and inadvertently created future fuel sources.

Cuts in appropriations to the Reinvest in Minnesota (RIM) program, the state's principal funding source for fire-management efforts, have resulted in fewer controlled burns. Unfortunately, these operations are expensive and labor intensive. The Minnesota DNR must rent equipment such as a helicopter to spread an igniting solution, bulldozers to build containment breaks, and ATVs to transport supplies to the fire team. And there is evidence to suggest that annual fires are necessary to control aspen. Infrequent fires may actually help the tree's progress. Aspen is a fire-adapted species, sprouting so vigorously after a blaze that it has been documented to produce seventy root suckers per square meter.

But people often hasten the succession of prairie to woodland in more direct ways. Some state and federal programs in the United States, for example, still encourage the planting of conifers on brush-lands, a practice that started with the public works programs in the 1930s. In other cases, consortiums of hunters have purchased large tracts of land in the Tallgrass Aspen Parkland. Believing that wood-

A species of willow (*Salix maccalliana*) unique to the Tallgrass Aspen Parklands. (Photo by Robert Dana, courtesy of the Minnesota Natural Heritage and Nongame Research Program)

lands increase game, most notably white-tailed deer, these groups have planted or allowed trees to flourish on their properties, even though the broken woodland habitat is preferred by deer. Finally, with the importance of aspen in a rapidly expanding forest-products industry may come a greater reluctance to control its growth in northwestern Minnesota, and an increasing desire to plant faster-growing hybrid poplars in and near prairie and brushland.

In spite of these problems, large portions of the tallgrass aspen parkland ecosystem still remain because poor drainage and marginal soils have prevented wholesale conversion of the land to crops. As a result, public agencies and private organizations have recognized the region's unparalleled opportunities for conservation. In Minnesota, large blocks of tallgrass aspen parkland and numerous small parcels have been set aside as private hunting preserves or as state wildlife management areas and parks. In Kittson County, for example, more than 121,000 acres of tallgrass aspen parkland remain in eight large blocks, including 59,000 acres in public ownership and 65,000 acres in private hands. The sizes of these nearly contiguous blocks range from 2,000 to nearly 60,000 acres in the Beaches-Skull Lake Parkland. Nearby Marshall County hosts about a dozen tallgrass aspen parkland remnants, and Pen-

Distant view to the southeast: aspen-parkland landscape with the Community Pasture in Manitoba and Caribou Wildlife Management Area in northwest Minnesota. (Photo by Gene Fortney)

nington County to the south contains more than 5,000 acres of contiguous tallgrass aspen parkland, with fifteen species of rare plants and animals, including the western prairie fringed orchid. Perhaps the best example of Minnesota's tallgrass aspen parkland ecosystem lies in and around the Caribou Wildlife Management Area in Kittson County near the Canadian border. This unfragmented patchwork of private and public lands provides an important complement to the nearby 12,000-acre Gardenton Community Pasture and the Tall-Grass Prairie Preserve, which protects more than 5,000 acres in Manitoba's Tolstoi-Gardenton area.

Together these acres begin to build a network of wild lands, which are rare in other agricultural regions. For example, the tallgrass aspen parkland provides habitat for the sharp-tailed grouse, which requires large blocks of land. In the United States, the sharp-tail has disappeared from most of its former range due to agricultural development, as well as poorly sited conifer plantations and the encroachment of the aspen forest. Woodlands are especially devastating since sharp-tails are not adapted to eating forest foods or spotting woodland predators, such as hawks and owls, which use the trees as hunting roosts.

The sharp-tailed grouse, so exquisitely fitted to the particular circumstances of the tallgrass aspen parkland, can be seen as a barometer of the ecosystem's vitality. Saving the sharp-tail, like preserving its unique home, depends as much on changes in biological management as it does on a shift in cultural attitudes, on our ability to give up—in this place at least—our love for trees and accept the land of the long horizon on its own terms.

Residents of the Tallgrass Aspen Parkland have created several model strategies for capitalizing on the region's unique natural resources. One of the most innovative approaches to aligning conservation and economic development goals can be found in Gardenton, Manitoba. Like many northern rural communities in recent decades, Gardenton, eighty miles south of Winnipeg, has experienced a steady loss in population and agricultural base. Also at risk was something

St. Michael's Ukrainian Orthodox Church, est. 1898 near Gardenton, Manitoba. (Photo by Gene Fortney)

less tangible but just as important—the area's rich ethnic heritage, which dates to 1896, when Ukrainian immigrants settled the area. Not as fertile as farther west, these pioneers still found valuable fuel and lumber in the aspen woodlands dotting the prairie. The settlers built a tiny town surrounded by churches, schools and farmsteads, including St. Michael's, Canada's first Ukrainian Greek Orthodox church.

97

Here Today, Gone Tomorrow?

Agricultural conversion and aspen encroachment have shrunk the populations of many open grassland bird species. Among the most troubling to Minnesota wildlife managers is the sharp-tailed grouse, whose numbers have plummeted in the state since 1980. Unlike its cousins the greater prairie-chicken, adapted to open grasslands, and the ruffed grouse of forested habitats, the sharp-tail's special habitat niche is the in-between place of open brushlands. During nesting and rearing periods, for example, both females and young offspring depend on the prairie and shrub thickets for cover as well as sustenance, such as insects, leaves, and seeds.

In Minnesota, the sharp-tail once occupied a range that extended from Albert Lea north to Cook County and west to the Dakota borders. On a journey up the Minnesota River in 1835, the English geologist George William Featherstonhaugh remarked that "the left bank of the river here was literally alive with [sharp-tails] coming to feed and drink from the burnt prairie; they were so large and fat, that they looked like barndoor fowl."

Even as recently as twenty years ago, the birds were so numerous that sharp-tails ranked third behind ruffed grouse and pheasants in the annual bird hunt. In the past five years, the population has dropped by seventy-seven percent in the bird's northwest and east-central ranges, the only two regions in Minnesota still inhabited by sharp-tails. According to Minnesota DNR research biologist Bill Berg, "We are standing by and watching sharp-tails disappear from Minnesota." (Photo by Craig Borck, courtesy of the *St. Paul Pioneer Press*)

Celebrating Prairie Day. (The Nature Conservancy photo archives)

Ukrainian Easter eggs. (Photo by Linda Shewchuk)

In 1965, Gardenton residents began to formally preserve their pioneer heritage through the development of the Ukrainian Museum, which is open during the summer months. Today, the town's annual Ukrainian festival in July draws thousands of visitors, who come to tour its Ukrainian museum, historic buildings, and farm fields, as well as sample the town's renowned ethnic cuisine—a stuffed dumpling known as a perohy ("pe-ro-gee").

For the past three years, the town of Gardenton has added another component to this heritage-tourism strategy. On the second weekend in August the town hosts Prairie Day, highlighting the nearby Manitoba Tall-Grass Prairie Preserve, whose three large land holdings, totaling more than 5,000 acres, form Canada's largest preserve of tallgrass prairie.

Gardenton has had a special longtime relationship with these tallgrass prairie lands. To minimize the dangers of wildfires, as well as to revitalize their fields, generations of farmers periodically set fire to their acres around the time of the Ukrainian Easter celebration. This practice benefited pastures and cropped fields, as well as patches of surrounding wild prairie that were left uncultivated because their soils were too poor, wet, or rocky to farm.

In 1989, the Manitoba Critical Wildlife Habitat Program began acquiring prairie lands in the Gardenton region. Today these remnants have become an important part of illuminating historic Ukrainian agricultural patterns and showcasing Manitoba's spectacular natural heritage. Linda Shewchuk, president and coordinator of the Ukrainian Museum and Village Society in Gardenton, observes that people are just beginning to learn about the prairie. In the future, the Gardenton tourism initiative hopes to capitalize on the community's natural resources by incorporating a hike through the prairie as part of its Ukrainian heritage bus tours. To generate additional visibility for the prairie, the town built an interpretive walking trail from the museum through the adjacent prairie. These tours, Shewchuk says, will emphasize the rare species of insects, birds, and plants that are unique to the area, which include Canada's only population of the endangered western prairie fringed orchid. At the same time, the nature walk will "link the cultural heritage to the prairie and help people understand the hardships faced by the homesteaders who settled here," Shewchuk observes.

Gardenton slowly is becoming the nucleus of a tourism hub to promote the human and natural history of the Tallgrass Aspen Parkland. In 1995, the town of Emerson, Manitoba, located twenty-five miles southwest of Gardenton and three miles north of Pembina, North Dakota, published a feasibility study for developing the nearby historic site of Fort Dufferin. Mayor Wayne Arseny explains that the town hopes to complement a new $2.2-million state of North Dakota museum in Pembina with the restoration of the fort, the site of several annual heritage events and recreational trails. According to the study, "heritage and tourism are seen as the industries of the future in the Emerson area." But the report stresses that Fort Dufferin should be viewed as one part of a greater round of area tourism destinations, adding that "it would be a great mistake to examine only the Fort

Linda Shewchuk (top) president and coordinator of the Ukrainian Museum and Village Society in Gardenton. Wayne Arseny (bottom) Mayor of Emerson, Manitoba. (Photos by Michael Burian)

Dufferin and Emerson areas. . . . The entire Emerson District and the Red River Corridor, from the Tallgrass Prairie to the east to the Aboriginal and the various 'ethnic' communities of the region with their cultural/heritage activities, all have roles to play in the development of what could be significant opportunities to enhance the entire region."

Communities across the border in Minnesota have not promoted ecotourism as actively as their counterparts in Manitoba, but "the potential is there, especially the nonconsumptive uses," says George Davis, area wildlife manager with the Minnesota DNR's office in Karlstad, Minnesota. Davis points to the region's unique and vast natural offerings. "These landscapes exist nowhere else in the world," he says.

Davis says that too often, however, public land holdings are landlocked by private property, limiting access and increasing the risks during fire management. The Minnesota chapter of The Nature Conservancy recently took an innovative approach to resolving this checkerboard dilemma. In northeastern Kittson County lies one of Minnesota's finest examples of tallgrass aspen

Checker-board of public ownership, Kittson County, Minnesota, around 1978. (Map courtesy of Minnesota Dept. of Natural Resources, Planning in cooperation with the Legislative Commission on Minnesota Resources)

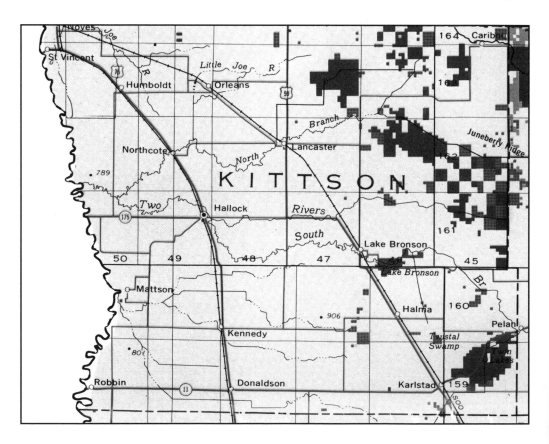

parkland. In 1995, the Conservancy purchased approximately 4,300 acres to consolidate the state's land holdings within the Caribou Wildlife Management Area (WMA). These acres were purchased at fair market value and, with the approval of the Kittson County government, transferred to Minnesota DNR ownership. Under the state's payment-in-lieu-of-taxes system, Kittson County will receive tax payments of three dollars per acre, more than is generated under present property-tax rates. Another 1,300 acres bordering the WMA were purchased for a Conservancy preserve or resale on the open market. Attached to the deeds of sold lands will be conservation easements, a popular land-conservation agreement. These buffers protect the core WMA by prohibiting land development on its periphery, at the same time leaving owners free to use their property within legally binding limits.

But there is growing recognition in the Tallgrass Aspen Parkland, as in other regions of the tallgrass prairie, that setting land aside is only one part of a comprehensive conservation strategy to preserve the region's unique resources. Several conservation groups and government agencies are collaborating with farmers and ranchers in the Tallgrass Aspen Parkland to develop sustainable practices. Under a federal program, the Canada and Manitoba Agreement on Agricultural Sustainability (CMAAS), the Manitoba Department of Agriculture is working with farmers to revitalize regional agriculture using native ecosystems. CMAAS supports more than seventy local networks of landowners interested in soil and water conservation projects. Its program has extended into the province's tallgrass prairie with the Stuartburn-Piney Agricultural Development Association, a rotational grazing demonstration project near Tolstoi. Rotational grazing allows farmers to move live-

Tallgrass Prairie Managment and Utilization Project near Tolstoi, Manitoba. (Photo by Gene Fortney)

stock through a series of paddocks on native grasslands, simulating the beneficial disturbances once caused by bison without damaging the land by overgrazing. Preliminary studies show that this grazing regimen improves water quality by protecting vegetation around wetlands and streams, encourages greater grassland biodiversity and increases the productivity of pastures, resulting in better-fed livestock.

The Manitoba Habitat Heritage Corporation also is looking at ways to help landowners reap economic, environmental, and recreational benefits from sustainably managing the woodlots on their properties. Through the Manitoba Agro-Woodlot Program, land-

Wheat harvest on Schwenzfeier farm, Hallock, Minnesota. (Photo by Don Breneman, courtesy of University of Minnesota Extension Service)

owners can access technical assistance for developing their own woodlot-management plans. The program is also investigating potential new markets for woodlot products in an effort to help diversify the economy of Manitoba's grain belt.

Similar sustainable agriculture initiatives could flourish in the United States. For example, it was proposed that the 1996 Farm Bill allow commercial uses to be incorporated into a revised CRP program in which farmers accept reduced CRP payments in exchange for limited opportunities to hay or graze CRP lands. Additional incentives, however, need to be developed for the region's small-grain farmers who lack either the know-how, capital, or the desire to raise livestock. Furthermore, in a region dominated by grain production, this alternative may also depend on strengthening or rebuilding the local economic infrastructure for supporting livestock production, such as auction barns, feedlots, and processors.

Diversifying the economy of the Tallgrass Aspen Parkland will, by extension, enrich the human and natural communities that sustain it. The task of striking a more habitable balance with our neighbors—whether songbirds, rare orchids, or fellow farmers—requires that we bring as many imaginations as possible to bear on repairing the frayed bonds that support life in the land between timber and tallgrass.

There's an aesthetic quality to this landscape that is by itself a strong reason for trying to hang on to it. For people who favor more dramatic ideas of landscape beauty, such as the Rockies or the sea coast, this land is stark and has a kind of lonely feel to it. I think it's gorgeous. To me it's a very northern landscape. When you're out there in the late summer and the sandhill cranes are overhead and the sun is raking across the land, it's incredibly beautiful.

Robert Dana, ecologist, Natural Heritage and Nongame Wildlife Program, Minnesota Department of Natural Resources

Appendix

Sustaining the Prairie and Parkland:
What Can You Do?

Many of us do not want to see the prairie and parkland disappear. We know it is a valuable resource—as important to us as our soil and water, as vital to us as our intelligence and energy, and as profound as the migration of the cranes, geese, and ducks through the Red River Valley every spring.

What can you as an individual do to anchor the prairie and parkland to this great place, keep it vibrant and full of life, and share its promise with unborn generations who will follow us? Here are some ideas to begin:

- Get to know prairie and parkland. Visit one on public land, at a nature preserve, or environmental learning center.
- If you own prairie and parkland, find ways to use it well. Do not plow it, build on it, or treat it in a way that drives its native species away. Find a way—as people telling their stories in this book have—to use prairie and parkland and keep it whole.
- If you want to sell your prairie or parkland, first contact an organization that will preserve it forever: a federal, state, or provincial wildlife group, a conservation organization, or a land protection trust.
- Let your municipal, county, and other local governments know about your support for prairie and parkland conservation. Support the efforts by organizations and people who are ensuring that prairie and parkland will survive in the valley for centuries to come.

Summer Range Tour in the Sheyenne Delta. (Photo by Jay Mar, Lake Agassiz Resource Conservation and Development Council)

107

To learn more, you can visit one of the many remarkable prairies and parkland areas open to the public:

In the Sheyenne Delta

The Sheyenne National Grassland (70,180 acres). The Sheyenne National Grassland is part of the 135,000-acre "Sandhills," which includes a deltaic plain, hummocky landforms, choppy hills, and river terraces. The deltaic plain supports tallgrass prairie, as do the tops of some hummocky landforms. Rare calcareous fens occur on river terrace slopes. The following five sites are included in the Sandhills.

Sheyenne Springs Research Natural Area, Ransom County (57.5 acres). This spring-fed wetland complex has a calcium-carbonate peatland, or fen, as its most extensive habitat and is surrounded by woodland. Sheyenne Springs is among the ten most significant wetland sites in North Dakota in terms of the number of rare species it contains and its undisturbed wetland habitats. **Directions:** Located forty-eight miles southwest of Fargo, ND, along the Sheyenne River. From Lisbon take State Highway 27 east for sixteen miles to County Road 53. Turn left (north) and go eight miles (the road bends east then north again) to a gated entry on the right side of County Road 53. Walk east one-half mile on the unimproved two-track road until you reach the tree line at the RNA's southwest edge.

Mirror Pool North State Natural Area, Ransom County (320 acres). Mirror Pool North is located north of the Sheyenne River. It spans more than a mile of the Sheyenne River and extends from the uplands to the floodplain forest. The dry, choppy dune habitat is a striking contrast to the sheltered forest and extensive wetlands. The mature forest represents some of the best eastern deciduous forest in North Dakota. **Directions:** From Lisbon, go sixteen miles east on State Highway 27 to County Road 53. Turn left (north) and go nine miles, then turn east and drive through a cattle guard. Travel by automobile beyond this point is not recommended. Walk southeast 1.25 miles to a second cattle guard at the west side of the state land.

Mirror Pool Swamp State Natural Area, Ransom County (151 acres). Located to the south of Mirror Pool North is a former oxbow of the Sheyenne River, which is the largest and one of the best-quality wetland basins in the state. Bordered on three sides by steep wooded slopes that grade into sand deposits, the swamp is fed by constant groundwater seepage and underlain by a peat deposit, which makes it a species-rich habitat.

Directions: From Lisbon, go sixteen miles east on State Highway 27 to County Road 53. Turn left (north) and go 7.5 miles, then turn right (east) and cross the Sheyenne River. Go 0.75 mile, then bend left and go 0.2 mile. Travel by automobile beyond this point is not recommended. Walk east on the two-track road 0.5 mile to the top of the hill, where the state land is noticeably posted.

Pigeon Point Preserve, Ransom County (560 acres). This preserve contains a series of spring-fed wetlands along the Sheyenne River valley margin, consisting of wetlands, forest, and prairie thickets. Wetland communities on this site include fens and numerous wetland thickets. Above the preserve is the historic site of the old Pigeon Point Trading Post. **Directions**: Located on the south side of the Sheyenne River. From Sheldon go seven miles south on County Road 54, turn left (east) and go five miles, then turn left (north) and go one mile to the northeast corner of the preserve, posted with Nature Conservancy signs.

North Country National Scenic Trail. This trail through the National Grasslands is twenty-six miles long. The National Grasslands begins at the Missouri River north of Bismarck and extends eastward through Minnesota, Wisconsin, Michigan, and Pennsylvania into northeastern New York. **Directions to western trailhead**: From Lisbon go ten miles east on State Highway 27 to County Road 54. Turn right (south) for three miles to the posted trailhead on the east side of the county road across from a gravel road. **Directions to eastern trailhead**: Go south from Leonard on State Highway 18. Turn east on State Highway 46 for 0.1 mile, then turn south on County Road 23, which has many turns. Travel four miles, then go east for one mile, then south 3.5 miles to the posted parking area for the trailhead.

In the Agassiz Beach Ridges

Foxhome Prairie, Wilkin County (240 acres). This low, wet, mesic tallgrass prairie and marsh, located along and below the basin edge of the highest relict beach of Glacial Lake Agassiz, is west of Fergus Falls and three miles northeast of Foxhome in Wilkin County. The eastern border of the preserve is Judicial Ditch-J2, built in 1906, which is the boundary between Wilkin and Otter Tail Counties. **Directions**: Take State Highway 210 between Foxhome and Breckenridge to County Road 19. Turn north and proceed three miles, then turn east on a gravel road. Follow this road for 1.5 miles; it will lead to Nature Conservancy signs along the south side of the road. Park on the north, east, or south side of the preserve.

Town Hall Prairie, Wilkin County (200 acres). This mesic and wet prairie is located on a belt of land nearly two miles wide, which slopes downward and lies parallel to the highest beach ridge of Glacial Lake Agassiz. It is 0.25 mile south of another Conservancy preserve, Anna Gronseth Prairie. **Directions**: From Rothsay: drive west on County Road 26 for approximately two miles until it intersects with County Road 19. Turn south and go four miles to County Road 20 and turn west; drive 0.5 mile until you see Nature Conservancy signs on the right side of the road; park alongside of the road.

Anna Gronseth Prairie, Wilkin County (1,221 acres). This low, wet, marshy tallgrass prairie preserve near Rothsay is a good representation of typical greater prairie chicken habitat. **Directions**: From Rothsay, travel west on County Road 26 for about five miles. Turn south on County Road 169. Travel about 3.25 miles, and park along the road on the west edge of the preserve.

Richard M. & Mathilde Rice Elliot Scientific and Natural Area (SNA), Wilkin County (497 acres). Found on this preserve is dry and mesic prairie, as well as sedge meadows and willow thickets. It is located between Barnesville and Fergus Falls. **Directions**: From Barnesville, go south four miles on County Road 52 to County Road 188. Turn east for one mile to the southwest corner of the preserve. Continue 0.5 mile to a pull-off on the south side of the road.

Western Prairie SNA, Wilkin County (320 acres). A tallgrass prairie preserve with rich alkaline soils, twenty-nine miles southeast of Moorhead in Wilkin County. Sedge meadow covers nearly fifty percent of the natural area, with wet and mesic prairies also present. In addition, a small well-defined saline area occurs in the SNA. **Directions**: Midway between Rothsay and Barnesville via State Highway 52. Take County Road 30 west of Lawndale for three miles, then go south on County Road 167 for two miles to the northwest corner of the preserve. Proceed 0.5 mile south to a pull-in on your left, next to the preserve sign.

Bluestem Prairie SNA, Clay County (3,258 acres). One of the largest and highest-quality northern tallgrass prairies in the United States. Included in the preserve are wet prairie, sedge meadow, mesic tallgrass prairie, dry prairie, and calcareous fens. This preserve lies fourteen miles east of Moorhead, adjacent to the Moorhead State University Science Center and Buffalo River State Park. Together, these three sites protect a 4,658-acre tallgrass prairie complex. It includes a 100-acre restored prairie, situated on land that was formerly ditched, farmed, and mined for gravel. **Directions**: Take U.S. Highway 10 at Moorhead east fourteen miles to State Highway 9. Turn south for 1.5 miles, then east on an unnumbered gravel road. Continue for 1.5 miles, and park on the north side of the road.

Margherita Preserve Audubon Prairie, Clay County (480 acres). Located in the lake bed of the ancient Glacial Lake Agassiz, this preserve includes the Herman beach ridge shoreline, found on the eastern edge of the preserve. This site is primarily wet and wet-mesic prairie. **Directions**: Take U.S. Highway 10 east from Moorhead for about twenty-five miles to Hawley. Turn right on County Road 31 and drive five miles south to Township Road 119. Turn right (west) and travel three miles to the center of the east boundary of the preserve. The last 0.25 mile has poor driving conditions.

Felton Prairie Complex, Clay County: Bicentennial Prairie SNA (160 acres), Blazing Star Prairie SNA (160 acres), and Shrike Unit SNA (90 acres). Located twenty miles northeast of Moorhead and four miles southeast of Felton, this is the most important gravel prairie complex in the state. Mesic black-soil prairie covers most of the site with substantial inclusions of gravel prairie on the beach ridges and wet prairie on the low swales. Many boulders left behind from the glacier are visible. **Directions**: From Moorhead take State Highway 9 north to two miles south of Felton and go two miles east on County Road 108 (a gravel road) for about 0.25 mile to the western boundary of the Bicentennial unit. The Shrike Unit is two miles east of Felton on County Road 34.

Zimmerman Prairie, Becker County (80 acres). This preserve, located thirty miles northeast of Moorhead, lies within the bed of a small glacial lake and is predominantly mesic tallgrass prairie. **Directions**: Take U.S. Highway 10 east two miles to State Highway 32 from Hawley to Ulen, to County Road 18 just past the Becker County line. Enter from the south boundary of the tract off Becker County Road 16.

Frenchman's Bluff SNA, Norman County (51 acres). Located five miles southeast of Twin Valley and thirty-five miles northeast of Moorhead. This area contains one of the highest points in northwestern Minnesota because of the formation of knobby hills and glacial till. **Directions**: From Twin Valley go south on State Highway 32 for about three miles, then turn left onto County Road 37 and continue east four miles. Turn right and drive south on County Road 36 about 0.75 mile to the top of the hill and look for Conservancy signs to your right. Park on the road shoulder.

Sandpiper Prairie SNA, Norman County (160 acres). This site is located about six miles west of Twin Valley and contains examples of dry, wet, and dry-mesic prairie communities. **Directions**: Go west out of Twin

Valley on County Road 27, travel 6.5 miles past the airport, and then go 1.5 miles south on the gravel road to the western boundary of the prairie. Park on the side of the road.

Prairie Smoke Dunes SNA, Norman County (780 acres). A dry-sand savanna community dominates this ancient inter-beach area near Twin Valley. Dunes range from forty to fifty feet high, with the leading edge of the dune forming an abrupt scarp nearly sixty feet high. This is a high-quality dune area, which also contains extensive prairie. **Directions**: From Twin Valley go fourteen miles north on State Highway 32, then 0.25 mile west on County Highway 7.

Agassiz Dunes SNA, Polk and Norman Counties (435 acres). Located near Fertile, it is within the largest dune field in Minnesota associated with Glacial Lake Agassiz. This area consists of dry sand savanna and dry sand prairie containing unique sand blowouts. **Directions**: From Fertile, drive south on State Highway 32, cross the Sand Hill River, and travel 0.6 mile. Turn west on a gravel road and continue for 0.5 mile. Turn south on a dirt road, which leads to a grass parking area. Do not drive onto the preserve.

Agassiz Environmental Learning Center, Polk County (640 acres). The Fertile Sand Hills offer spectacular overlooks, riverine forest, and dunes; in addition, parts of the Sand Hill River remain open all winter, providing wildlife and birding opportunities. **Directions**: The center is located just outside of Fertile, which is fifteen miles south of U.S. Highway 2 on State Highway 32.

Malmberg Prairie SNA, Polk County (80 acres). This prairie stands out as virtually the last relict of prairie on the flat Red River Valley proper. **Directions:** In Crookston, drive west on County Road 9 from where it intersects with U.S. Highway 75. Turn left onto County Road 56 and continue two miles to reach the northwest corner of the preserve.

Pankratz Memorial Prairie, Polk County (North Unit: 322 acres; South Unit: 456 acres). Situated on a Glacial Lake Agassiz beach ridge, the land is largely wet-mesic prairie with low swales and, on the south unit, a thirty-five-acre calcareous fen. **Directions to North Unit**: From Crookston drive east on U.S. Highway 2 for six miles. Turn south on County Road 46 and travel 1.5 miles to reach the northwest corner of the unit. To park, turn east on the gravel road by the preserve sign. **Directions to South Unit**: Drive 1.5 miles south from the preserve sign and turn east and go 0.5 mile to the west side of this section of the preserve.

Pembina Trail Preserve SNA, Polk County: Pembina Trail Unit (1,657 acres), Crookston Prairie Unit (560 acres), and Foxboro Prairie Unit (160 acres). These tallgrass prairies are some of the most spectacular prairies in northwestern Minnesota. They lie on the west edge of a conspicuous Glacial Lake Agassiz beach ridge; granite boulders, which are scattered throughout, served as bison rubbing rocks more than a century ago. **Directions:** Southeast of Crookston take State Highway 102 to the intersection with County Road 45. Drive east for 6.5 miles and watch for the signs. Park at the entrance of an old field road leading into the preserve.

In the Tallgrass Aspen Parkland of Minnesota

Caribou Wildlife Management Area (WMA), Kittson County (13,500 acres, including 1,340 acres of School Trust land). These lands abut the Canadian border beginning approximately seven miles east of U.S. Highway 59. They are a large, intact example of tallgrass aspen parkland into which naturally occurring elk have migrated. **Directions:** Access is difficult, but best achieved from County Road 4. From Lancaster, travel east on County Road 4 for twenty-one miles (road bends north, then east, then north again.) State land—the east end of the WMA—is west of the road.

Beaches WMA, Kittson County (30,325 acres, including 5,000 acres of School Trust land). This is the largest continuous block of tallgrass aspen parkland in the United States, with a wide variety of habitats and wildlife. The eastern half of this area is mostly wetlands. **Directions:** Access is difficult. From Lancaster travel east six miles on County Road 4 to where the road turns north. Excellent tallgrass aspen parkland runs along the east side of this road for two miles, but state land lies one mile farther east of the private land here.

Skull Lake WMA, Kittson County (7,480 acres, including 1,120 acres of School Trust land). A large example of tallgrass aspen parkland. **Directions:** Access is difficult. From the Canadian border, travel 2.5 miles south on U.S. Highway 59. Turn east and go three miles. The northwest corner of the WMA is one mile south.

Lake Bronson State Park, Kittson County (3,000 acres). This is an easily accessible example of tallgrass aspen parkland. **Directions:** From Lake Bronson on U.S. Highway 59, travel one mile east on County Road 28 to the state park.

Norway Dunes, Kittson County (320 acres). This unusual, but small, nature preserve contains steeply sloping sand dunes formed by wind and covered with oak savanna. **Directions:** From Halma travel east on County Road 7 for 0.5 mile, turn north on a township road for two miles. The preserve entrance is 0.25 mile east on a field road.

In the Tallgrass Aspen Parkland of Manitoba

Manitoba Tallgrass Prairie Preserve, Rural Municipality of Stuartburn (5,000 acres). This area contains surviving remnants of the original tallgrass prairie, together with aspen groves and large areas of wetland. Most lands in the preserve are accessible year-round for hiking or walking; however, vehicle access may be limited. **Directions**: Travel eight kilometers (five miles) north of the U.S.-Canadian border on Provincial Trunk Highway 59 to Tolstoi. Travel east on Provincial Road 209 toward Gardenton. The public prairie lands begin about 2.5 kilometers (1.5 miles) east of Tolstoi.

Lake Francis Wildlife Management Area, Rural Municipality of Woodlands (16,000 acres). This area contains about 2,000 acres of wet to wet-mesic tallgrass prairie and also includes marshes and aspen bluffs. The beach front along Lake Francis is an important nesting site for birds. **Directions**: From Provincial Trunk Highway 101 in Winnipeg, travel northwest thirty-two kilometers (nineteen miles) on Highway 6. At Provincial Road 411, go west about sixteen kilometers (ten miles) to the wildlife management area.

Oak Hammock Wildlife Management Area, Rural Municipality of Rockwood (280 acres). Two blocks of tallgrass prairie, roughly 200 acres and eighty acres respectively, lie within this WMA. This sixty-five-square-mile expanse of marsh, meadow, tallgrass prairie, willow shrub, and aspen-oak bluffs is a reclaimed remnant of St. Andrews Bog, a large habitat that almost disappeared before its restoration was undertaken. This is also a renowned spring and fall staging area for an estimated 400,000 geese and ducks. **Directions**: From Winnipeg, travel north eighteen kilometers (eleven miles) on Provincial Trunk Highway 7 to Provincial Trunk Highway 67. Go east eight kilometers (five miles) to Provincial Road 220; turn north and go three kilometers (two miles) to the interpretive center.

Living Prairie Museum, Winnipeg (30 acres). This tallgrass prairie contains more than 160 species of native plants. Two large boulders, known to have been "bison rubbing stones," mark the east side of the nature center. Special programs feature native performers, storytelling, traditional crafts, bison burgers, bannock, nature hikes, and children's games and activities. Nature hikes at Assiniboine Forest, Bradley Prairie, LaBarriere Park, and Little Mountain Park, all within or adjacent to Winnipeg, are also available. **Directions**: Located in the city of Winnipeg at 2795 Ness Avenue.

For more detailed written information about many of these sites, please contact:

In the Sheyenne Delta: U.S. Forest Service, Sheyenne Ranger District, P.O. Box 946, Lisbon, ND 58054 (701-683-4342); The Nature Conservancy of North Dakota, 2000 Schafer St., Suite B, Bismarck, ND 58501 (701-222-8464).

In the Agassiz Beach Ridges: The Nature Conservancy, Northern Tallgrass Prairie Office, Route 2, Box 240, Glyndon, MN 56547 (218-498-2679); Scientific and Natural Areas Program, Minnesota Dept. of Natural Resources, 500 Lafayette Road, Box 7, St. Paul, MN 55155 (612-296-3344).

In the Tallgrass Aspen Parkland: Critical Wildlife Habitat Program, c/o Wildlife Branch, Manitoba Dept. of Natural Resources, Box 24, 200 Salteaux Crescent, Winnipeg, MB R3J 3W3 (204-945-7750; Manitoba Naturalists Society, 302-128 James Ave., Winnipeg, MB R3D 0N8 (204-943-9029); Living Prairie Museum, 2795 Ness Ave., Winnipeg MB R3J 3S4 (204-832-0167); Wildlife Division, Minnesota Dept. of Natural Resources, P.O. Box 154, Karlstad, MN 56732 (218-436-2427); The Nature Conservancy of Canada, Manitoba Office, 298 Garry St., Winnipeg, MB R3C 1H3 (204-942-6156).

Pasque Flower (*Anemone patens*) lifts purple heads in April, a sign of resurrection on the prairie. (The Nature Conservancy photo archives)

Selected Resource Organizations

Prairie and Parkland Region

Glacial Lake Agassiz Interbeach Area
 Stewardship Team
c/o Peter Buesseler
Minnesota Department of Natural Resources
1221 E. Fir Avenue
Fergus Falls, MN 56537
(218) 739-7497

Great Plains Partnership
Web Site:http://www.eerc.und.nodak.
edu/rrbin/gpp/gpphome.html

The International Coalition
P.O. Box 127
Moorhead, MN 56561
(218) 233-0292

Red River Basin Information Network
Web Site: http://www.eerc.und.nodak.edu/rrbin

Sheyenne Delta

Lake Agassiz Resource Conservation &
 Development Council
417 Main Avenue
Fargo, ND 58103
(701) 239-5373

The Nature Conservancy
2000 Schafer Street, Suite B

Bismarck, ND 58501
(701) 222-8464

North Dakota Department of Game and Fish
Rt. 1, Box 224
Jamestown, ND 58401
(701) 252-4681

Sheyenne Grazing Association
15 Main Street
McLeod, ND 58057
(701) 439-2670

USDA Natural Resources Conservation Service
P.O. Box 1458
Bismarck, ND 58502-1458
701-250-4425

U.S. Forest Service
Sheyenne Ranger District
Box 946, 701 Main Street
Lisbon, ND 58054
(701) 683-4342

Agassiz Beach Ridges

Agassiz Environmental Learning Center
P.O. Box 388
Fertile, MN 56540
(218) 945-3204

Clay County Planning Office
807 11th St.
Moorhead, MN 56560
(218) 299-5041

The Nature Conservancy
Northern Tallgrass Prairie Office
Route 2, Box 240
Glyndon, MN 56547
(218) 498-2679

Minnesota Department of Natural Resources
Minerals Division
2115 Birchmont Beach Road NE
Bemidji, MN 56601
(218) 755-4067

Minnesota Department of Natural Resources
Natural Heritage and Nongame Research
Program
500 Lafayette Road, Box 7
St. Paul, MN 55155
(612) 297-4964

Minnesota Department of Natural Resources
Scientific and Natural Areas Program
1221 E. Fir Ave.
Fergus Falls, MN 56537
(218) 739-7497

Resource Conservation and Development Assn.
Pembina Trail
2605 Wheat Drive
Red Lake Falls, MN 56715

U.S. Fish and Wildlife Service
Detroit Lakes Wildlife Management District
Route 3, Box 47D
Detroit Lakes, MN 56501
(218) 847-4431

U.S. Fish and Wildlife Service
Fergus Falls Wildlife Management District
Highway 210 East
Fergus Falls, MN 56537
(218) 739-2291

Tallgrass Aspen Parkland, Minnesota

Minnesota Department of Natural Resources
P.O. Box 154
Karlstad, MN 56731
(218) 436-2427

Tallgrass Aspen Parkland, Manitoba

Critical Wildlife Habitat Program
c/o Wildlife Branch
Manitoba Department of Natural Resources
Box 24, 200 Saulteaux Crescent
Winnipeg, MB R3J 3W3
(204) 945-7750

Manitoba Conservation Data Centre
    Wildlife Branch
Manitoba Department of Natural Resources
Box 24, 200 Saulteaux Crescent
Winnipeg, Manitoba R3J 3W3
(204) 945-7743

The Nature Conservancy of Canada
Manitoba Office
298 Garry Street
Winnipeg, Manitoba RC3 183
(204) 942-6156

Manitoba Naturalists Society
302-128 James Avenue
Winnipeg, Manitoba R3D 0N8
(204) 943-9029

Living Prairie Museum
2795 Ness Avenue
Winnipeg, Manitoba R3J 3S4
(204) 832-0167

Prairie Habitats
Box 1
Argyle, Manitoba R0C 0B0
(204) 467-9371

Manitoba Museum of Man and Nature
190 Rupert Avenue
Winnipeg, Manitoba R3B 0N2
(204) 956-2830

Selected Sources of
Information

Chapter 1. Our Common Ground

Campbell, M. 1872. Letter to his sister Isabella. Clay County Historical Society, Moorhead, Minnesota.

Hind, H.Y. 1971. *Narrative of the Canadian Red River Exploring Expedition of 1857 and of the Assiniboine and Saskatchewan Exploring Expedition of 1858.* M.G. Hurtig, Ltd., Edmonton.

Krenz, G., and J. Leitch. 1993. *A River Runs North: Managing an International River.* Red River Water Resources Council.

Minnesota Prairie Chicken Society. 1985. *The Prairie Chicken in Minnesota.* Minnesota Prairie Chicken Society, Crookston, Minnesota.

Spry, I. 1963. *The Palliser Expedition: An account of John Palliser's British North American Expedition, 1857-1860.* Toronto, Macmillan Co. of Canada.

Chapters 2 & 3. Thinking like a Prairie

Apfelbaum, S.I., J.D. Eppich, T.H. Price, and M. Sands. In press. The Prairie Crossing Project: Attaining water quality and stormwater management goals in a conservation development. Pages 33-38 in *Proceedings of a National Symposium: Using Ecological Restoration to Meet Clean Water Act Goals,* March 14-16, 1995. U.S. Environmental Protection Agency, Chicago.

Bennett, H. 1948. The tools of flood control. In *Grass: The Yearbook of Agriculture.* U.S. Department of Agriculture, Washington, D.C.

Hartnett, D.C., A.A. Steuter, and K.R. Hickman. 1997. "Comparative Ecology of Native and Introduced Ungulates." Pages 72-101 in F.L. Knopf and F.B. Samson, eds., *Ecology and Conservation of Great Plains Vertebrates.* Ecological Studies, vol. 125. Springer, New York.

Hopkins, D. 1997. "Hydrologic and Abiotic Constraints on Soil Genesis and Natural Vegetation Patterns in the Sandhills of North Dakota." Ph.D. dissertation, North Dakota State University, Fargo.

Swerman, R., D. Baker, and R. Skaggs. 1987. *Minnesota Drought.* Water Resources Research Center, University of Minnesota, St. Paul.

Taff, S.J. 1989. *The Conservation Reserve Program in Minnesota: 1986-89 Enrollment Characteristics and Program Impacts.* University of Minnesota, Minnesota Agricultural Experiment Station, Minnesota Report 217-1989.

Tilman, D., and J.A. Downing. 1994. "Biodiversity and Stability in Grasslands." *Nature* 367: 363-65.

Tyler, V.E. 1994. *Herbs of Choice: The Therapeutic Use of Phytomedicinals.* Pharmaceutical Products Press, Haworth Press, New York.

U.S. Army Corps of Engineers and Minnesota Department of Natural Resources. 1996. Environmental impact study of flood control impoundments in northwestern Minnesota: Final Environmental Impact Statement. U.S. Army Corps of Engineers and Minnesota Department of Natural Resources, St. Paul.

Weaver, J.E. 1954. *North American Prairie.* University of Nebraska Press, Lincoln, Nebraska.

Chapters 4 - 6. Living with the Prairie

Berg, W.E. 1990. "Sharp-tailed Grouse Management Problems in the Great Lakes States: Does the Sharp-tail Have a Future?" *The Loon* 62: 42-45.

Breining, G. 1996. "Roads and Ridges." *Minnesota Volunteer* 59 (Jan.-Feb.): 46-51.

Duram, L.A. 1995. "National Grasslands: Past, Present and Future Land Management Issues." *Rangelands* 17(2): 36-41.

Elson, J.A. 1967. "Geology of Glacial Lake Agassiz." In W. Mayar-Oakes, ed., *Life, Land, and Water.* University of Manitoba Press, Winnipeg.

Murray, S.N. 1967. *The Valley Comes of Age: A History of Agriculture in the Valley of the Red River of the North, 1812-1920.* North Dakota Institute for Regional Studies, Fargo, North Dakota.

Nelson, W.T., W.T. Barker, and H. Goetz. *Habitat Type Classification of Grasslands of Sheyenne National Grassland of Southeastern North Dakota.* U.S. Forest Service, Report for Cooperative Agreement No. RM-80-19-CA.

State of Minnesota. 1997. Clay County Beach Ridges Forum: Final Report. Minnesota Department of Natural Resources. St. Paul.

Recommended Reading

Allman, L. 1996. *Land Protection Options: A Handbook for Minnesota Landowners.* The Nature Conservancy, The Trust for Public Lands, Minnesota Land Trust, and Minnesota Department of Natural Resources Natural Heritage Program and Scientific and Natural Areas Program, Minneapolis and St. Paul.

Anonymous. *Sustainable Development: An Action Plan for a Network of Special Places for Manitoba.* 1992. Sustainable Development Coordination Unit, Province of Manitoba, Winnipeg.

Clambey, G.K. and R.H. Pemble. 1986. *The Prairie—Past, Present and Future: Proceedings of the Ninth North American Prairie Conference.* Tri-College University Center for Environmental Studies, Moorhead, Minnesota.

Critical Wildlife Habitat Program. 1996. *A Guide to Conservation Programs and Funding Sources for Agri-Manitoba: Land Stewardship Directory.* Critical Wildlife Habitat Program, Manitoba Department of Natural Resources, Winnipeg.

Higgins, K.F. 1986. *Interpretation and Compendium of Historical Fire Accounts in the Northern Great Plains.* U.S. Department of Interior, Fish and Wildlife Service, Resources Publication 161, Washington, D.C.

Higgins, K.F., A.D. Kruse, and J.L. Piehl. 1986. *Effects of Fire in the Northern Great Plains.* State Cooperative Fish and Wildlife Research Unit, Extension Circular 761, University of South Dakota, Brookings.

The International Coalition. 1989. *Land and Water Guide: Red River Basin.* The International Coalition, Moorhead.

Ladd, Doug. 1995. *Tallgrass Prairie Wildflowers.* Falcon Press, Helena, Montana.

Minnesota Audubon Council. 1993. *Minnesota Wetlands: A Primer on Their Nature and Function.* Minnesota Audubon Council and National Audubon Society, St. Paul.

Minnesota Interagency Exotic Species Task Force. 1991. *Report and Recommendations to the Natural Resources Committees of the Minnesota House and Senate, April 1991.* Minnesota Department of Natural Resources, St. Paul.

Minnesota Natural Heritage Program. 1993. *Minnesota's Native Vegetation: A Key to Natural Communities.* Biological Report No. 20. Minnesota Department of Natural Resources, St. Paul.

Minnesota Scientific and Natural Areas Program. 1995. *A Guide to Minnesota's Scientific and Natural Areas.* Minnesota Department of Natural Resources, St. Paul.

Moffat, M., and N. McPhillips. 1993. *Management for Butterflies in the Northern Great Plains: A Literature Review and Guidebook for Land Managers.* U.S. Fish and Wildlife Service, Pierre, North Dakota.

Morgan, J.P., D.R. Collicutt, and J.D. Thompson. 1995. *Restoring Canada's Native Prairies: A Practical Manual.* Prairie Habitats, Argyle, Manitoba.

Ostlie, W.R., and T.M. Faust. 1997. *An Assessment of Biodiversity in the Lake Agassiz Interbeach Area: An Ecoregion within the Great Plains.* The Nature Conservancy, Great Plains Program, Minneapolis.

Stoner, J.D., D.L. Lorenz, G.J. Wiche, and R.M. Goldstein. 1993. "Red River of the North Basin, Minnesota, North Dakota, and South Dakota." *Water Resources Bulletin* 29: 575-615.

Tester, J.R. 1995. *Minnesota's Natural Heritage: An Ecological Perspective.* University of Minnesota Press, Minneapolis.

Tornes, L.E., and M.E. Grigham. 1994. *Nutrients, Suspended Sediment, and Pesticides in the Waters of the Red River of the North Basin, Minnesota, North Dakota, and South Dakota, 1970-1990.* U.S. Geological Survey, Water-Resources Investigations Report 93-4231.

Trottier, G.C. 1992. *Conservation of Canadian Prairie Grasslands: A Landowner's Guide.* Environment Canada, Canadian Wildlife Service, Edmonton.

Wedin, D.A. 1992. "Grasslands: A Common Challenge." *Restoration and Management Notes* 10: 137-43.